Science and Fiction

For further volumes:
http://www.springer.com/series/11657

Science and Fiction – A Springer Series

This collection of entertaining and thought-provoking books will appeal equally to science buffs, scientists and science-fiction fans. It was born out of the recognition that scientific discovery and the creation of plausible fictional scenarios are often two sides of the same coin. Each relies on an understanding of the way the world works, coupled with the imaginative ability to invent new or alternative explanations - and even other worlds. Authored by practicing scientists as well as writers of hard science fiction, these books explore and exploit the borderlands between accepted science and its fictional counterpart. Uncovering mutual influences, promoting fruitful interaction, narrating and analyzing fictional scenarios, together they serve as a reaction vessel for inspired new ideas in science, technology, and beyond.

Whether fiction, fact, or forever undecidable: the Springer Series "Science and Fiction" intends to go where no one has gone before!

Its largely non-technical books take several different approaches. Journey with their authors as they

* Indulge in science speculation – describing intriguing, plausible yet unproven ideas;
* Exploit science fiction for educational purposes and as a means of promoting critical thinking;
* Explore the interplay of science and science fiction – throughout the history of the genre and looking ahead;
* Delve into related topics including, but not limited to: science as a creative process, the limits of science, interplay of literature and knowledge;
* Tell fictional short stories built around well-defined scientific ideas, with a supplement summarizing the science underlying the plot.

Readers can look forward to a broad range of topics, as intriguing as they are important. Here just a few by way of illustration:

* Time travel, superluminal travel, wormholes, teleportation
* Extraterrestrial intelligence and alien civilizations
* Artificial intelligence, planetary brains, the universe as a computer, simulated worlds
* Non-anthropocentric viewpoints
* Synthetic biology, genetic engineering, developing nanotechnologies
* Eco/infrastructure/meteorite-impact disaster scenarios
* Future scenarios, transhumanism, posthumanism, intelligence explosion
* Virtual worlds, cyberspace dramas
* Consciousness and mind manipulation

Michael Brotherton

Editor

Science Fiction by Scientists

An Anthology of Short Stories

 Springer

Editor
Michael Brotherton
Dept. 3905, University of Wyoming
Laramie, Wyoming, USA

ISSN 2197-1188   ISSN 2197-1196   (electronic)
Science and Fiction
ISBN 978-3-319-41101-9   ISBN 978-3-319-41102-6   (eBook)
DOI 10.1007/978-3-319-41102-6

Library of Congress Control Number: 2016955554

Cover illustrations: Front cover: Man standing on top of the hill watching the stars, illustration painting, © HYPERLINK "http://www.shutterstock.com/gallery-2352014p1.html" Tithi Luadthong. Back cover: Photo by John Gilbey.

Printed on acid-free paper

This Springer imprint is published by Springer Nature
The registered company is Springer International Publishing AG
The registered company address is: Gewerbestrasse 11, 6330 Cham, Switzerland

Preface

I love science.

I love science fiction.

Since I was a kid, science and science fiction have been two sides of the same coin. At age six I was watching Star Trek and begging to go to the Natural History Museum and their awesome dinosaurs at every opportunity. Amazing creatures from the distant past and exotic worlds from the distant reaches of the galaxy, these were things that either science or science fiction could bring me, but nothing else could.

Science and the technology it spawns changes the world, bringing us knowledge, space, and the future itself. Well done science fiction provides a glimpse into realistic and amazing futures – or terrible futures we as a society should avoid. With the appeal of the wonders of the universe, and the bonus of foreseeing avoidable disasters, I could not stay away. I pursued my twin loves throughout my life, eventually becoming an astronomy professor who also wrote science fiction novels steeped in astrophysics.

To me, the distinguishing element of science fiction has always been and always shall be the "science" part, but there is plenty of "science fiction" on bookshelves and the movie screens that has precious little to do with science. Without the science, it's just a western in space, or maybe a fantasy set in the future. There are audiences for those, and that's fine. There are writers who aspire to deliver the science, but find it difficult, and that's fine, too. Luckily I was not the first, nor the last, to become both a scientist and a science fiction writer.

Scientists can deliver on the science, and there is a history of delivery on the fiction as well. Isaac Asimov earned a PhD in chemistry before turning to writing full time and creating the three laws of robotics and the psychohistory of his Foundation trilogy. Arthur C. Clarke brought us *2001: A Space Odyssey*, and also was the first to link geostationary orbits to electronic communications. Fred Hoyle coined the term "The Big Bang Theory" (derisively, to be fair), and his thrilling speculation gave us the sentient space gas of *The Black Cloud*. Physicist Robert Forward's brilliant imagination brought us a vision of life on the surface of a neutron star in *Dragon's Egg*, as well as serious proposals for laser-propelled sails to voyage to other stars. Carl Sagan's best-selling novel *Contact* about a positive SETI result also spawned a successful Hollywood blockbuster. Gregory Benford, a physics professor, won the Nebula award for his 1988 novel *Timescape* that realistically depicted not only tachyons but the academic world of science. There are many dozens of other scientists who write science fiction, coming from increasingly diverse disciplines and backgrounds, such as David Brin, Catherine Asaro, Vernor Vinge, Alastair Reynolds, and Geoffrey Landis.

This collection highlights a new generation of twenty-first century scientist science fiction writers. The majority are active research scientists, working at universities, medical schools, and space agencies, drawn to write stories on the side. Others are full-time writers who have retired from science, or, like Asimov, have set aside a career in science to write. In addition to the more traditional astronomers and physicists, the contributors include biologists, neuroscientists, computer scientists, and rocket scientists.

Given the technical expertise of these contributors, we have taken advantage of the opportunity to get them to further discuss the science in their stories in afterwords following each contribution. As one Star Trek character might opine about the far-out science explored in these pages, "fascinating."

I still love science and science fiction as much as when I was a kid, and I hope you'll find these tales as fascinating as I do.

Laramie, WY   Michael Brotherton

Biographical Sketches
of Authors

Jed Brody teaches physics at Emory University. As a participant in the Emory-Tibet Science Initiative, he traveled to India five times to teach physics to Tibetan monks and nuns. He was a Peace Corps volunteer in Benin, West Africa. He is the author of two science-fiction novels, *The Philodendrist Heresy* and *The Entropy Heresy*. 100% of his royalties from sales of these novels are donated to charity.

Eric Choi is an aerospace engineer and award-winning writer and editor based in Toronto, Canada. He holds a bachelor's degree in engineering science and a master's degree in aerospace engineering, both from the University of Toronto, and he is an alumnus of the International Space University. Over the course of his engineering career, he has worked on a number of space projects including QEYSSat (Quantum Encryption and Science Satellite), the MET (Meteorology) payload on the Phoenix Mars Lander, the MSS (Mobile Servicing System) robotics on the International Space Station, the RADARSAT-1 Earth-observation satellite, and the MOPITT (Measurements of Pollution in the Troposphere) instrument on the Terra satellite. In 2009, he was one of the Top 40 finalists (out of 5351 applicants) in the Canadian Space Agency's astronaut recruitment campaign. He was the creator and co-editor of two speculative fiction anthologies, *Carbide Tipped Pens* (Tor) with Ben Bova and *The Dragon and the Stars* (DAW) with Derwin Mak. The first recipient of the Isaac Asimov Award (now the Dell Magazines Award) for his novelette "Dedication", he is also a two-time winner of the Prix Aurora Award – the Canadian national prize for excellence in speculative fiction – for his short story "Crimson Sky" and for co-editing *The Dragon and the Stars*. Please visit his website www.aerospacewriter.ca or follow him on Twitter@AerospaceWriter.

Andrew Fraknoi is the Chair of the Astronomy Department at Foothill College near San Francisco, and was the California Professor of the Year in 2007. With the late Byron Preiss, he co-edited *The Universe and The Planets*, two anthologies of science fact and fiction published in the 1980s. He is also the lead author on an introductory astronomy textbook, *Voyages through the Universe*, and wrote a book for children, *Disney's Wonderful World of Space*. He keeps a reading list of science fiction featuring reasonable astronomy at: www.astrosociety.org/scifi. Fraknoi was the Executive Director of the Astronomical Society of the Pacific for 14 years, and serves on the Board of the SETI Institute and on the Lick Observatory Council. Asteroid 4859 was named Asteroid Fraknoi by the International Astronomical Union in recognition of his work in public education, but he is eager to reassure readers that it is a well-behaved main-belt asteroid, and poses no danger to Earth.

Carl(ton) Frederick is a theoretical physicist, at least theoretically. After a post-doc at NASA he did a stint at Cornell University. There, he wrote a paper on Stochastic Space-time that some considered groundbreaking. Nonetheless, he became disillusioned with academia and left his first love, research on the fundamentals of quantum theory (a strange first love, perhaps) and succumbed to the enticements of hi-tech industry. He invented the, now totally obsolete, 1200 baud digital modem, and Venture Capital moved him and his company, Wolfdata, to Boston. Soon though, tired of being a lance-corporal of industry, he left his company and moved back home to become Chief Scientist of a small group doing AI software. While keeping his hand lightly in theoretical physics, he decided he'd like to write a more overt form of science fiction and, to that end, enrolled in the Odyssey Writers Workshop. He subsequently earned a first place in Writers of the Future. He now has a respectable corpus of published short-stories including 45 sales to *Analog*. He has put an interactive novel on the Web. It is interactive in that you can click to change the point of view and to expose sub-plots (www.darkzoo.net should you care to visit). He's written a half dozen or so novels and, after shopping them around faster than a speeding glacier, has turned them into Kindle e-books where they are now, along with numerous collections of his short stories, moldering in obscurity on Amazon. (You can find them by searching on Amazon for 'Frithrik', his college nickname.) He has two grown children and shares his house with a cat and a pet robot. For recreation, he fences épée, learns languages, and plays the bagpipes. He lives in rural, Ithaca, New York. And rural is good if you play the bagpipes. He has since returned to his aforementioned first love.

Les Johnson is a physicist and the Technical Advisor for NASA's Advanced Concepts Office at the Marshall Space Flight Center where he serves as the Principal Investigator for the NASA Near-Earth Asteroid Scout solar sail mission. Les is an author of several popular science books including *Solar Sails: A Novel Approach to Interplanetary Travel* [featured in *Nature*, April 2008] and *Harvesting Space for a Greener Earth*. He is also a science fiction writer; his books include *Going Interstellar*, *Rescue Mode*, and 2016's, *On to the Asteroid*. Les was the featured 'interstellar explorer' in the January 2013 issue of National Geographic magazine. He thrice received NASA's Exceptional Achievement Medal and has three patents. To learn more about Les, please visit his website at www.lesjohnsonauthor.com.

Edward M. Lerner has degrees in physics, computer science, and business administration. He worked in high tech and aerospace for thirty years, as everything from engineer to senior vice president, for much of that time writing science fiction as his hobby. Since 2004 he has written full-time. His novels range from near-future technothrillers, like *Small Miracles* and *Energized*, to traditional SF, like his InterstellarNet series, to (collaborating with Larry Niven) the space-opera epic Fleet of Worlds series of *Ringworld* companion novels. Lerner's most recent novel, *InterstellarNet: Enigma*, won the inaugural Canopus Award "honoring excellence in interstellar writing." His fiction has also been nominated for Locus, Prometheus, and Hugo awards. Lerner's short fiction has appeared in anthologies, collections, and many of the usual SF magazines. He also writes about science and technology, most notably in his long-running "The Science Behind the Fiction" series of essays for *Analog*.

Marissa Lingen is a science fiction writer living in the Minneapolis suburbs. She has published over one hundred short stories in venues such as *Nature, Analog, Tor.com, Twenty-First Century Science Fiction*, and several Year's Best anthologies. Before becoming a full-time writer, she studied physics at Gustavus Adolphus College, University of California-Davis, and Lawrence Livermore National Labs. She did research projects in interstellar spectroscopy and ceramics before settling on a nuclear physics focus to her graduate work but decided that writing was a better fit. She hikes when she can, bakes when she can't, and makes paper art inspired by neurons.

Stephanie Osborn, the Interstellar Woman of Mystery, is a 20+-year space program veteran, with graduate/undergraduate degrees in astronomy, physics, chemistry and mathematics, is "fluent" in several more, including geology and anatomy. She has authored, co-authored, or contributed to over 25 books, including the celebrated novel, *Burnout: The mystery of Space Shuttle STS-281*. Co-author of the *Cresperian Saga*, she currently writes the critically-acclaimed *Displaced Detective Series*, described as "Sherlock Holmes meets The X-Files," and the new *Gentleman Aegis Series*. She "pays it forward," teaching STEM through numerous media including radio, podcasting and public speaking, and working with SIGMA, the science-fiction think tank.

Jon Richards is a Senior Software Engineer at the SETI Institute concentrating on detecting SETI signals using the Allen Telescope Array. He is a computer engineer comfortable developing in many programming languages and many different types of computer systems. His past work has involved a lot of hardware design and development, tying hardware and software to networks and the internet. Since 2008 he has been trying to continually build his skills and knowledge of digital signal processing and trying to master the Allen Telescope Array hardware and software. For more information Jon and his work, see http://www.seti.org/users/jrichards

Tedd Roberts is the pseudonym of neuroscience researcher Robert E. Hampson, Ph.D. For more than 35 years, he has studied physiology & pharmacology, learning & memory, and brain impairment in many forms (head injury, epilepsy, abused drugs and radiation). He is involved in a research collaboration to develop a "neural prosthetic" for restoring human memory function. A keen interest in public education and brain awareness led him to join the National Academy of Science's Science and Entertainment Exchange, provide subject matter expertise to SF/F writers and game developers, and to write science fact articles and science fiction stories of his own. With more than 150 professional research articles, he chooses to publish his nonfiction 'Science-in-Science Fiction' articles and SF short stories under his pen name "so that my colleagues can tell the difference!" Dr. Hampson is a medical school professor, married for over 30 years, with two grown sons. In between travel, teaching, speaking, studying martial arts and playing trombone in a Brass Octet, he makes his home in the Piedmont region of North Carolina.

Jennifer Rohn is Principal Research Associate in the Division of Medicine at University College London, United Kingdom. She has B.A. in Biology from Oberlin College, Ohio and a Ph.D. in Microbiology from the University of Washington in Seattle. She has been involved in cell, molecular and microbiological research in both academic and biotech settings in several different countries since 1989, and currently heads a research team investigating the subversive molecular behavior of the bacteria involved in chronic urinary tract infection. Jennifer also has a long-standing interest in the portrayal of scientists in fiction. She coined the term "lab lit' and founded the popular science/culture website LabLit.com to encourage more writers to use science and scientists in their everyday fiction. She has written two novels, *Experimental Heart* and *The Honest Look*, both published by Cold Spring Harbor Laboratory Press and loosely inspired by her experiences in biology laboratories over the years. Her short fiction has appeared in *Nature* and *The Human Genre Project*.

J.M. Sidorova holds a Ph.D. in molecular genetics and she is a faculty member of the University of Washington School of Medicine, where she studies DNA replication in normal and cancerous human cells. J.M.'s science fiction and fantasy short stories appeared in *Clarkesworld*, *Asimov's*, *Abyss and Apex*, and other venues. Her debut novel *The Age of Ice* (Simon & Schuster), nine parts history, one part magic realism, was featured on *Locus Magazine*'s recommended reading list, and received an honorable mention on *Tor.com*'s best fiction of 2013 list. As a translator, she contributed to the *Red Star Tales*, an anthology of Russian science fiction (Russian Life Books, 2015). She is a graduate of the Clarion West workshop. She can be found online at www.jmsidorova.com.

Ken Wharton has been a physics professor at San Jose State University since 2001. His research is in Quantum Foundations, a field that strives for a deeper account of quantum theory and a better understanding of what quantum phenomena might be telling us about our universe. (A general-level essay describing Dr. Wharton's overall research program can be found online under the title "The Universe is not a Computer".) His 2001 novel *Divine Intervention* (Ace) was awarded the Special Citation for the Philip K. Dick award, and he has also been a finalist for both the Nebula and the Campbell Awards.

J. Craig Wheeler is the Samuel T. and Fern Yanagisawa Regents Professor of Astronomy, Distinguished Teaching Professor at the University of Texas at Austin, and past Chair of the Department. He has published nearly 300 refereed scientific papers, as many meeting proceedings, a popular book on supernovae and gamma-ray bursts (*Cosmic Catastrophes*), two novels (*The Krone Experiment* and *Krone Ascending*), and has edited six books. He co-wrote a screenplay of *The Krone Experiment* with his son, Rob, that Rob subsequently made into an independent microbudget film. Wheeler has received many awards for his teaching, including the Regents Award, and is a popular science lecturer. He was a visiting fellow at the Joint Institute for Laboratory Astrophysics (JILA), the Japan Society for the Promotion of Science, and a Fulbright Fellow in Italy. He has served on a number of agency advisory committees, including those for the National Science Foundation, the National Aeronautics and Space Administration, and the National Research Council. He has held many positions in the American Astronomical Society and was President of the Society from 2006 to 2008. He currently serves on the AAS Ebooks committee. His research interests include supernovae, black holes and astrobiology.

Contents

Down and Out

Ken Wharton

Ogby trudged up the seamount, expanding her bladders as forcefully as she could, but the effort didn't gain her much weight. Her body was becoming so light it felt like the current was going to sweep her away, footholds or no footholds.

The surrounding spectrum shifted oddly for a moment. Ogby paused in confusion until she saw three lampfish swimming just above her head, altering the artificial light patterns on the icy slope. She jealously watched the fish swim against the current. The biologists were now claiming that Rygors must have once been able to maneuver like fish, way back in their own evolutionary past. But her more recent ancestors had forgotten how to swim, spending their lives pinned to the bottom of the ocean by the bladders in each of their five feet. And while swimming might have been useful at these elevations, apparently her ancestors never had a need to come up this high. Or perhaps, considered Ogby, they had been petrified of being swept upward to their deaths.

She cautiously peered to the left to see how high they had come, and was struck by a vicious wave of vertigo. The city lights at the bottom of the seamount looked impossibly far away. Expanding her bladders helped fight the sensation, but not much; her muscles were weak after spending so much time in the Deeps. She closed her eyes and forced herself to draw in a long, continuous jet of water through her funnels. *The feeling will pass*, she told herself.

© The Author 2017
M. Brotherton (ed.), *Science Fiction by Scientists*, Science and Fiction,
DOI 10.1007/978-3-319-41102-6_1

By the time she opened her eyes, the others had stopped ahead to wait for her. "I heard she was afraid of heights," Roov was chroming to no one in particular.

Ogby flashed the group an apologetic pattern, while simultaneously soning for them to "GO AHEAD." She was embarrassed to have slowed down the whole group, but they refused to move on until Ogby resumed her climb.

After another five milliflexes of hiking, Ogby finally joined the others at the top of the seamount. Her feet were tender and sore from stretching her bladders, but she had made it to the Boarding Station.

Roov was clearly not having the same troubles — he even let go of the footholds and performed a little hop to show his lack of fear. Ogby wondered who he was showing off for. Vyrv, perhaps? But Vyrv was already in the ship, beckoning the rest of them to enter.

Ogby tipped back her head and looked up at the cable, stretching from the top of the ship into the darkness above. She was worried. If she had been afraid of heights on the mount, how would she feel, suspended underneath the very roof of the world?

Intellectually, she knew it would be safe. She would be inside the entire time, at a controlled pressure. And even if the cable snapped, the ship had an active buoyancy control. But her fear was stronger than her logic, and a sudden wave of fresh panic nearly kept her from entering the ship.

In the end it was her scientific curiosity that won. The interesting research was happening Above. If she wanted to participate in the latest discoveries, she would have to conquer her fears. She grimly stepped inside the ship to join the others.

The workers closed the hatch, locking in the water pressure for the remainder of the journey. As Ogby stretched her sore fingers, one foot at a time, she noted that the cabin interior was almost identical to the ships she piloted down in the Deeps. On one side were the primary controls: wheels and levers that controlled the compressed air tanks to regulate the ship's buoyancy. In the center were the cylindrical passenger benches, with those new plastic seat-covers made from greenfish oil. Ogby straddled a bench and strapped herself in. The other passengers did the same, all except Roov who took the control seat.

"I always insist on piloting the ship myself," chromed Roov to the others. "Just in case there's an emergency."

Ogby tried not to show her exasperation. Roov was full of himself, but he was also one of the most influential scientists in the ocean. His discovery Above of the new element "gold" had made him famous with the average citizen, and he had been able to use his clout to funnel additional money into the overhead research and mining efforts. If it hadn't been for Roov's tacit approval, Ogby wouldn't be here right now.

A sudden lurch, and then the ship was in motion. Ogby averted her gaze from the windows; the sight of the Boarding Station dropping away beneath her would do little to calm her nerves.

The altimeter needle on the control panel was rising rapidly; they were already a full kilolength above standard ground. Six more and they'd be at the top of the ocean.

"This is your first time up, too?" Vyrv asked her.

"Yes," chromed Ogby. "I've spent a lot of time in these ships, but never way up here."

"Oh?" Vyrv seemed surprised. "Where, then? Down in the Deeps? Didn't think there was much down there. Just ice."

"There has to be *something*," Ogby insisted. "Whirlpools must go *somewhere*."

Roov joined his colors into the conversation, ignoring the controls now that the counterweight was lifting them at the proper speed. "Whirlpools are an anomaly; everyone knows that ice is heavier than water. The way I see it, natural causation moves downward, with us Rygors the ultimate consequence at the bottom. Think about it. We eat the fish, which in turn eat the microscopic life, which in turn feed off the vents we've found Above. But what powers the vents? What's above the Above? Why does the ground flex in such a predictable rhythm? When we get to the top I'll show you the new excavation; we've dug higher up into the rock than ever before. I'm sure that one day we'll break through to Outside, find out that our ocean is just a small part of a much bigger universe."

"You believe in Outside?" Vryv asked wryly.

"There must be an Outside," chromed Roov in all seriousness. "Yrvo's voyage proved that you can drift around the world, proved the ocean is a spherical shell. *Something* has to be outside."

"Not necessarily," flashed Ogby, hoping she wasn't being too impertinent. "For all we know, the rock up there goes out to infinity."

Roov turned his full attention in her direction, and paused before responding. "Instead of trying to disparage our work, you might take a look at your own. You've been digging in the ice for a kiloflex, and what have you discovered?"

Ogby didn't respond. In all of her Deep excavations, she had found precious little of interest. All of the major new discoveries had been made Above: the new elements, the new lifeforms, the Vents, the bubble factories. Below she had found only ice.

"I'm not disparaging you," Ogby insisted. "I would like very much to join your team."

"If so," chromed Roov, "the first thing you're going to have to do is prove you can handle the height."

Roov's colors dimmed, and little else was discussed for the remainder of the journey. Eventually the ship lurched to a halt. They had arrived at the top of the ocean.

After docking with the main habitat, the hatch opened and warm water diffused into the cabin. This was a curious fact no one had yet explained, Ogby knew. Up here the water was slightly warmer than down below. Yet the super-hot water from the Vents was heavy and carried the nutrients straight down to the bottom of the ocean. It didn't make sense to her, but then again, a lot of things about gravity didn't make sense.

Ogby was the second passenger to step out into the cylindrical walkway. The corridors were thinly air-cushioned; not so deep that she couldn't get traction, but still more comfortable than a solid floor.

Roov began the tour when everyone had left the ship. "Over here," he chromed, "are the intake valves. Specially designed to keep the water fresh without changing the interior pressure. But I'm sure you'll be more interested in the Vents. Come this way."

As Ogby approached the observation deck she had a premonition of disaster. Yes, she was interested in the Vents, but somehow she hadn't considered that in order to see outside of the habitat there must be windows. And with windows, she might look *down*. There would be no pretending that she was in a structure at the bottom of the ocean; her tremendous height was about to become very obvious. The thought made her fingers twitch in nervous anticipation.

And the reality was even worse. Instead of simply a room with glass portholes in the walls, the floor was also covered with small windows. She forced her attention upward before stepping in.

The observation deck was a circular platform built next to a particularly large Vent. The Vent itself looked like a narrow upside-down seamount, made out of rock instead of ice. Ogby kept her gaze high, examining the less-interesting upper portions of the Vent. Streaks of color told most of the geological story; some sort of material had sprayed out of the bottom of the Vent and then oozed up the sides before solidifying.

But the others were all looking through the floor, filling the room with color as they chromed their appreciation. Reluctantly, curiously, Ogby lowered her gaze.

It was a fantastic display. Superhot squirts of water pulsed regularly from the opening, so hot that they glowed in the far red. The surrounding water was also quite warm; a faint glow surrounded the entire bottom half of the Vent.

Ogby had never seen natural light before. To her, all light came from animals, Rygors, or Rygor-made objects like sonoluminescent lamps. On some primal level she felt the natural beacon summoning her, just as it must have summoned the creatures that teemed in the red glow. There were no familiar deep-water fish, but plenty of new species: a fish with far more fins than seemed necessary, another organism shaped like a slow-moving net, even a little 5-legged cutie which looked almost like a miniature Rygor.

This was where life started, she knew. This was where she needed to be. Up here she could find the answers she was looking for, figure out how the world worked. Down below lay only....

Down below.

Ogby couldn't help herself, and once she looked down it was impossible to stop. There were tiny lights down there, she saw, swimming against the black background. Black, because the bright lights from the cities couldn't reach these heights.

The distance hit her all at once. *I'm too high*, she thought. *I'm too high.*

Now the others were trying to talk to her, trying to get her to respond, but she didn't dare move. She wanted more than anything to get back to the ship, to get back to the ground, but she couldn't even walk off the deck.

She dimly realized she was being carried somewhere, with her eyes closed. Still, the fear wouldn't stop. "WAKE," someone soned at her, the sound reverberating painfully from the habitat walls. She felt herself shutting down, ancient survival mechanisms having their way with her body. At last her consciousness drifted deeper than even the bottom of the ocean, and all was dark.

***

"You've got to get back out there," Boro insisted, back in Ogby's underground web three flexes later.

Ogby watched her mate disinterestedly, wondering if she'd even keep him for another season. "What does it matter?" she chromed dimly. "I had one chance. I blew it. Roov won't let me try again."

Boro shook his middle legs before continuing. "I'm not telling you to get back *up* there. Just get back to work. There's plenty of interesting science you can do down here. The techs in the factory have been asking about you since yesterflex."

"I don't want to do science anymore," she responded. "I just want to be left alone."

"So you're through? You can't go Above, so instead you're just going to quit everything?" Boro turned away from her, but continued to chrome from his back. "What about your pressure calculations? I know you still think there's something under the ice."

"NO," she soned at him, but Boro didn't even turn around. In fact, now he was leaving, just like she had asked. She almost soned him to STOP, but her pride kept her quiet, and soon he was out of the web completely.

Still, maybe Boro was right. After seeing the splendor of the Vents and the mysteries they contained, she had forgotten about the more mundane problems she studied down here.

The physics had been known even to the ancients. A flexible bladder of air would change its size depending on elevation, and that in turn would change its weight. The fact that bladder size was proportional to weight had been known for hundreds of generations, possibly even megaflexes. But only recently, using the new excavators, had anyone been able to measure the effect deep *below* ground level.

Ogby herself had spearheaded the largest excavation yet, melting a kilolength deep into the ice. Roov was correct that she hadn't discovered anything down there, but she *had* discovered that the gravity continued to rise, even deep underground. And when she extrapolated the curve, it looked like gravity should go to infinity just 2.8 kilolengths below standard ground.

According to most other scientists, this was nonsense. Infinities were mathematical, not real. Yes, the ocean was a spherical shell, so they admitted something odd might happen down at the very center. But based on the calculations from Yrvo's round-the-world voyage, the distance to the center should have been megalengths, not kilolengths. No, the other scientists insisted, the change in gravity must slow with depth.

But despite the soundness of their logic, Ogby's numbers had shown no such trend. The only way to test it, she knew, was to dig down to minus 2.8 kl and see what happened. But at the rate she was going, it would take more than her lifetime to get that far.

Although… what had Roov said about the excavations Above? They were digging up there, too, but that was rock. And you couldn't melt through rock.

A milliflex later she was out of her web. Boro was already long gone, so she began cantering towards town. A new rain of bubbles had just fallen; pools of methane and carbon dioxide lay in the low valleys. She deflated her bladders and skidded across a small airpool, enjoying the smooth sensation on her feet.

Soon she came to the largest airpocket in the city. This was the primary local factory, much larger than any of her personal airlabs, but similar in concept. It had taken forever to dig the huge hole, let alone fill it up with carbon dioxide, but the goods produced here had already paid back its cost many times over. This factory specialized in plastics and steel, and trade with other cities brought in more exotic items.

With a long-practiced move she stepped onto the enormous bubble, inflated her bladders to maximum, and then *inverted* herself. It was a sensation that bothered many Rygors, but it felt perfectly natural to Ogby. Fully inflated, her bladders were so heavy they acted as anchors from which she could pull herself downward. Her ankles flexed 180°, and then she was dangling upside-down inside of the air pocket, barely touching the water with her five feet. The trick was mental reorientation: she told herself that she was actually standing right-side-up, her feet floating on the surface of the water. It was a ridiculous image, but it enabled her to avoid the unpleasantness experienced by many of the others.

It *was* unpleasant not being able to breathe, but Ogby was better than most at holding her breath. Only six or seven times each work period did she have to duck her head in the water tanks. And being in the air pocket allowed for other benefits. She began to refresh the stale air from her bladders, first from her central cavity, and then one foot at a time. It felt good.

But she still had to find Boro. Walking along the surface of the water Ogby presently arrived at the smallest forge, where Boro spent much of his time. She climbed up the stairs, opened the hatch, and there he was, just closing the thermal shielding around the primary steel cauldron.

"I'm surprised to see you," Boro chromed.

Ogby skipped the small talk. "Those excavations Roov is doing Above. He's not melting through, like we do. He's actually digging?"

Boro rippled a 'no'. "Blasting, I think. You know about those new compounds they're making, over in High City? I think he's using those, setting them off from a distance."

Ogby stood stunned for a moment. She had thought explosives were still in the research phase. Roov was already putting them to use? Science was progressing so fast these days that she couldn't even seem to keep up.

"Well," she said at last, "why can't we use them, too? I'm sure it would speed up the dig. Maybe we could even blast down far enough to test my calculations."

"Do you have any idea how much those things cost?" Boro replied.

"I don't need many to start with. Let's buy a few, give it a try in Deep 4."

Boro looked concerned. "If you really want to blow the last of your research money.... Oh sure, why not. I'll pilot the ship again, if you'd like."

"I'm piloting," Ogby chromed with a literal flash of defiance.

"You haven't piloted since we broke 800 lengths. You've just had a minor breakdown, and—"

"I can handle it, Boro."

Boro stared at her for a long time before responding. "Okay. I think I believe you."

Ogby flashed a contented pattern, then turned to leave. A strange noise made her stop, though, and when she looked back around she saw that Boro was soning her through the air. She was surprised; soning in air was incredibly painful. If he had simply wanted to get her attention....

"I wanted to tell you," Boro chromed, "it's good to have you back."

Ogby felt a sudden wave of attraction for Boro, the first such wave in many flexes. Was it her season already? She checked her specialized fingers on her third foot, somehow already knowing what she would find.

"What is it?" Boro asked.

Ogby wiggled her third foot at him enticingly. "I don't think we've ever done it in the air before."

He didn't look pleased. "Not now, Ogby."

Ogby was stunned. How could he resist…? But of course. There was no water to carry her scent. She walked over to him, reached for his third foot with her own, and made the transfer directly.

Boro put up no further resistance — not until he ran out of breath and desperately leaped into the emergency water tank. Ogby followed him right on in.

***

Deep 4 was the largest excavation in the ocean, far from the nearest city. Here the ice naturally dipped to half a kilolength below standard ground, and the entire valley would have been below air if not for constant maintenance.

Currently most of the crew was cowering in the generator shelter, but Ogby wanted to be outside when the package landed. Boro stood next to her nervously, along with three of the braver technicians. The lip of the circular excavation was just a few lengths away.

"It should have reached the bottom by now," chromed a tech. "I don't—"

At that moment the blast hit. Even with the pads over Ogby's sonar receptors, even with the explosion over a kilolength below, it felt like a series of body-blows.

Boro shuddered after the waves had passed. "Next time I'll be in the shelter," he chromed unhappily.

Ogby shook off the sensation and pushed Boro toward the edge. She was worried that the explosion would propel fragments of ice up in their direction. "Look down, see if there's any debris coming this way. I'm getting into the ship."

Now that the blast had arrived, every microflex counted. It was cold in the deeps, and only artificial heaters kept anchor ice from filling in the hole. Ogby had to get down there quickly if she was going to wire up the new heaters.

She started to close the ship's hatch, making sure everyone was at their stations. A final glance at Boro confirmed that no debris was going to endanger her. Thanking him, she shut herself inside and ran a cursory check of the equipment. Fuel, batteries... check. The comm light was on, but she never trusted it blindly. She activated the microphone with her first foot.

"TEST," she soned into the mike.

She looked out of the port window at the giant spool of cable, 2.3 kilolengths long, which connected her ship to the shelter. Fortunately the sound was converted to electrical signals, or else the communication would have been unbearably slow. After a moment the words "test received" sounded from the ship's speaker. The cable was operational.

Now came the scary part; going over the edge. Ogby positioned herself in front of the controls and began adjusting the ship's buoyancy. After lifting off the ice, it only took a single thrust to position herself directly over the hole. It was a long way down, Ogby knew, but that was exactly where she had to go.

"DROPPING," she soned into the mike, while simultaneously shifting the plunger controls to negative buoyancy. In a moment she was plummeting into the cold, watery depths.

"FIRST HEATERS OKAY," she told base control as she passed the glowing devices. Aiming the ship's outer lights, she saw that the powercord was still firmly attached to the walls of the pit. Everything looked fine. She tipped the spotlight downward and continued her descent.

The view out of the lower window was the first indication that something was wrong. The bottom of the pit was still beyond the power of the ship's lights, but instead of trailing away into darkness, the depths suddenly turned a foamy white. And the whiteness was rising, fast.

Uh oh, was all Ogby had time to think before the first jolt hit the ship.

She was tossed to one side, and her head collided painfully with the cabin wall. After a moment the acceleration stopped, and Ogby quickly strapped herself into her seat.

What's happening? she asked herself. Debris from the explosion? No, any debris would have arrived with the original blast. It must be a second explosion, she decided, but how was that possible? They had only dropped one package—

Another, stronger jolt shook the ship, but the straps held. Then another, and another, and Ogby began to worry about the ship coming apart at the welds.

The view out the window offered little information. A dark froth of water and ice swirled past meaninglessly. "HELP," she soned, hoping the mike could pick up her voice from across the cabin. "EMERGENCY."

The buffeting continued, for ages, but just as she thought she could stand it no longer, cabin stopped shuddering. Now the window lit up with a brilliance she had never imagined possible. She narrowed her eyes, averted her head, but light was too strong, too painful.

And then, with a massive jerk, came the largest jolt of all. Ogby felt the straps cut into her body, and a tiny 'ping' sounded from the speaker just as gravity turned itself off.

The light from the window was slightly more bearable now, but she barely noticed. Down here gravity wasn't infinite; it was zero! The bladders in her feet felt no force at all; it didn't matter whether she clenched them or not. She didn't know how this was possible, but it was the discovery of a lifetime. She looked up toward the microphone…

…and her heart broke. The light was off. That 'ping' noise; it must have been the cable snapping. Communication was now impossible. Whatever was going to happen to her, whatever she encountered, she now had to face it alone.

Movement through the window caught her eye and she stared through it, amazed. The scene was bright, but no longer too bright. There was no water. Instead she saw a beautiful icy landscape, covered with fractures and lines. The colors were remarkable; new minerals shouted to her with their unique spectra, arrayed in branching linear patterns. And the whole landscape was growing, filling the window with its details, coming closer and closer…

A sudden crunch of metal, a terrible pain, and all went black.

***

Ogby regained consciousness slowly, vaguely aware that the water around her was cold. Too cold. Ice was already beginning to form in the upper corners of the ship.

She almost drifted back to sleep, almost content to die to this way. Then her curiosity got the better of her.

Even before unstrapping herself, she noted that gravity was pulling her to the bottom of the cabin with just as much force as ever. Perhaps the zero-gravity moment had only been a dream.

The pain, however, was real. Wincing as she moved her bruised body, she stepped over to check the controls. The batteries still had some power left, and the heaters were nearly on full. So why was the water so cold? She maxed out the heaters and moved over to the window.

The outside view was so strange that it took her a long time to parse it into something she could understand. The ship apparently lay on the underside of a giant, bright, icy plain; she had probably crashed into it from below. The ship must be buoyant here, she realized, as if she was inside a giant air pocket.

Her mind reeled with questions. Had she penetrated the ice, broken through to the center of the ocean? Why was there no water here? Why was it so bright? What was this place?

Closer to the ship, she saw that the icy plain was scarred, rippled in a circular pattern. And into the center of the ripples snaked a black line, coiling around itself until it disappeared into the ice.

The cable!

Her mind started to piece together a story. Something had sucked her down through the ice. Her ship had come flying through, launched down into the air pocket, stretching the cable tight. The cabin had jolted as the cable had snapped. Then the ship had reversed direction, and crashed upwards into the ice. But why? None of this made any sense to her.

Still, that cable… that was her link to a world that made sense. If she could reconnect it to the ship, she could call for help. There might even be enough slack to reach. But the only way to do it was to go outside, and the only way to go outside was to open the hatch, spilling out most of her water. She'd never survive long enough to be rescued.

Still, the ice in the cabin was continuing to spread — if she didn't act soon she'd be frozen solid, and no one would ever know what had become of her. Making a run for the cable would surely be better than that.

Steeling her resolve, she began to hyperfunnelate, readying her system for what lay ahead. She knew that the area outside must be very cold — the evidence was quickly crystallizing all around her — so she spent a few moments hunting for something to wear on her feet. But the only free objects were the plastic seat-cover slabs. She grabbed three of them with three feet, hoping they'd provide enough insulation.

Ice had already started to form on the hatch, but she broke it free with the plastic slabs and started turning the primary release wheel with her two free feet. One turn, two turns…

Without warning the door flew outward, pulling Ogby with it. Water spurted out around her, erupting into a frenzied boil. Panicked, she clung to the wheel, feeling the water rush past her as it left the ship.

She had been wrong about the temperature. If the water was hot enough to boil, she would be roasted alive in seconds.

But even as the water frothed around her, she noticed that it was quickly freezing onto the surface of the icy plain. Could it be cold after all? Her body was becoming uncomfortable, first aching all over, and now flaring with pain. Then she noticed that her air bladders were inflating on their own accord, and she had to clench them tightly to keep them under control.

Gravity wasn't infinite here, she suddenly realized. But the pressure was very low — maybe even zero. Zero pressure meant infinite volume, and her own bladders were struggling to obey that particular law of physics. Ogby recalled the evacuated chambers from the airlabs, and quickly guessed that she had just entered an enormous vacuum.

The water from the ship had now emptied completely. A quick glance inside told her that everything else had frozen solid. There would be no reserve to breathe later. And if she was in a vacuum, she knew there wasn't much time. She had to get to that cable quickly, before her body could no longer contain the pressure inside of her.

It wasn't far to the cable, only about thirty lengths. And now that she was out of water, the weak gravity pulled her upwards, directly toward the ice. She glanced in the other direction…

The view below nearly made her faint.

Directly beneath her, visible now that she was out of the ship, sat an enormous brilliant sphere, banded with swirling colors as if it were chroming to her in another language. A smaller yellow sphere floated on top of it, casting a shadow on the large sphere's misty surface. The large sphere also had a second shadow, although Ogby couldn't tell what was casting it. There must be an even brighter light source somewhere else....

Distractions, she told herself. She had to concentrate on the cable if she was to survive. Still, those spheres looked so far away that her vertigo was starting to kick in.

With a flash of insight she decided to try the reorientation trick she used in the factory. *That's not down*, she told herself, still staring at the spheres. *That's up. The ice is down.*

She positioned her legs accordingly, touched the three plastic slabs to the ice, and let go of the wheel. In a moment she was trotting across the surface towards the free end of the cable.

Yes! she mentally cheered herself on, imagining the world turned upside-down. *I'm running on the ice, on top of the ice, on the outside…*

Outside. Even as she approached the cable she knew it was true. She wasn't inside the ocean at all. The light source casting those shadows wasn't visible, so it had to be on the other side of the ice. The second shadow might even be her entire world!

She picked up the cable, her mind so busy that she barely appreciated the connector was still intact.

If she was Outside, then her entire worldview must be wrong. What she considered "up" was actually toward the center of her world; Roov's excavations through the rock could only break through to the other side of the ocean. The Cities were on the *outer* edge of the ocean, and the Rygors spent their lives with their feet pointed outwards and their heads pointed to the center. Yes, the ocean was a spherical shell, but it wasn't curved in the way that everyone had assumed.

Ogby was almost back to the ship now, her body ablaze with pain. The plastic slabs were already freezing in her grip, but she wouldn't let them go. She pulled more of the slack in her direction, shifting the cable connector in her grip, and at that moment her second foot exploded.

Agony flooded through her nervous system. Her airbladder was broken, in tatters, and she watched numbly as her juices quickly oozed out of the wound. Some corner of her mind knew that this was the end, this was where her internal pressure would equalize with the vacuum of this alien space. She had been so close…

But she wasn't dead yet. Ignoring the pain, she clenched the fingers on her second foot, applying pressure to the wound as best she could. She felt her second leg shudder, but the oozing slowed to a halt.

Somehow she reached the ship, plugged in the cable, climbed through the hatch. Her first foot did all the work, locking it shut behind her and moving over to the buoyancy controls.

Ogby needed pressure fast, and the tanks of compressed air were her only hope. Normally the air was released into the flexible buoyancy chamber, not the main cabin, but a few flipped levers rerouted the valves. Even as she felt her life slipping out of her second leg, her soning receptors began to pick up the unique noise of hissing air.

She rested for a moment as the pressure built up, feeling the pain in her foot subside. There was ice in the bottom of the cabin, but she knew she would never breathe again. All she had left to accomplish in this life was to communicate her findings down below.

Wincing with the effort, she pressed her receptors directly over the speaker. The device was designed to sone through water, not air, but her body seemed to work as a decent medium. Immediately she heard the faint words:

"orange, green, blue, yellow. speak. orange, green, blue, yellow. please speak."

They were calling her, spelling out the colors of her name as best they could via sound.

She soned a reply, expecting it to hurt more than it did. Compared to the rest of her injuries, the sensation of soning through air felt almost pleasant. Their response came quickly.

"received. received. where you? whirlpool here."

A whirlpool? The explosion in the hole had triggered a whirlpool! It all made sense to her now; whirlpools were simply the sucking of water from the high-pressure ocean, through the ice, to the zero-pressure of Outside.

She began to explain the situation to whomever was listening. It would have been a difficult explanation even if she could chrome, let alone using the paltry soning vocabulary, but somehow she managed to convey the basic story: she had discovered Outside.

She told them that for the next expedition, they should bring pressure suits, heaters, thermal shielding, light filters, cameras and recording equipment. She told them to watch the colored spheres and figure out what they could tell the Rygors about the rest of the universe.

It was only at the very end that she realized she had been soning with Boro for the entire conversation. She couldn't hear any emotion in his electronic voice, but somehow it still came through when he said goodbye.

Ogby somehow found the strength to sone one last sentence.

"WE MEET WHEN YOU COME OUTSIDE."

Leaning back on the remnants of her frozen water, Ogby's gaze through the window fell upon the large sphere that she had just discovered, far below. *Was the sphere chroming to her?* she dimly wondered. *Telling her all the secrets of the universe?* She tried to focus, but the last shreds of her attention could only note the sphere's most prominent feature, gazing back at her. When Ogby's consciousness finally slipped away, her final mental image was of that great, red spot.

Afterword

The appearance of the great red spot at the end of this story is supposed to indicate the setting. Specifically, this story takes place in the interior of Europa: an icy moon of Jupiter with a large ocean under the ice. Despite the very low surface temperatures, Europa's ocean does not freeze solid thanks to "tidal heating" from a slightly-eccentric orbit around Jupiter (in turn due to another moon, Io). The slight deviations in the distance between Europa and Jupiter result in a cyclic compression of Europa, with the same frequency as Europa's orbit. Europa's orbital period is the duration of the unit "flex" used in the story, as these compressions would be measurable from inside the moon.

The Rygors are an advanced lifeform living in Europa's ocean, very roughly modeled off octopi. The food chain is powered by hydrothermal vents at the bottom of the ocean, similar to known deep-sea ecosystems on Earth. But the

Rygors are not neutrally-buoyant like most terrestrial ocean creatures. Instead, each of their five arms has a small air bladder, which pulls them up to the top of the ocean, on the underside of Europa's crust-ice layer. (Presumably the Rygors use some biochemical reaction to produce a gas that can fill these bladders.) There are no hydrothermal vents up at the ice, so all of their food supply has to come via fish-like organisms that swim in the ocean.

The story is told from the Rygors' perspective: the underside of the ice is their "ground", and the buoyancy force from their air bladders defines the direction "down". From their perspective they walk on the "ground", and are pulled to it by "gravity"; from our outside-Europa perspective they are simply buoyant.

This inversion of our usual perspective makes for some unusual situations. We humans don't have a fear of being swept upwards to our death if we climb a high mountain. But this what induces Ogby's 'fear of heights'; not falling, but being swept "upwards" if she gets too "high".

The analogous concern from our perspective would be a scuba diver who gets too deep. Divers wear a Buoyancy Compensator device, or "BC", that can be inflated or deflated as desired. If a diver did not properly regulate the BC, descending into regions of higher pressure causes the BC volume to be compressed by the surrounding water. This would make it less buoyant — in principle sending an unwitting diver to the bottom of the ocean. (Archimedes' Principle tells us that the upward buoyancy force on an object is equal to the weight of the displaced fluid.) As the diver's BC gets compressed, it displaces less water, and becomes less buoyant. Flip this picture upside down, and this scenario is exactly what Ogby feared in the opening scene: that she would be unable to physically counter the exterior pressure, and she would find her air bladders compressed to the point where they provided very little buoyancy. From her perspective, buoyancy is gravity, so the gravity diminished as she climbed the mountain.

Hopefully the above explanation is enough physics background to follow some of the other curious aspects of the story. At one point I toyed with the idea of making this the beginning of a much longer story, allowing Ogby to be rescued on the surface of Europa by human visitors. The present version doesn't say what happens next, one way or the other, so it's not *necessarily* a sad ending. It just requires your imagination to continue the tale.

The Tree of Life

Jennifer Rohn

She would never forget the day they came. Although it wouldn't be difficult to remember, as "never" would only last, in the end, a few short months. This was clear from the outset. After that, she wouldn't need to remember anything, ever again.

She was working late, the last one in the lab. It was summer, though you wouldn't know it from the careful chill of the air, set at a humidity-controlled, regulation-standard 18 degrees centigrade. She was focused on the work spread out in front of her: a rack of small Eppendorf tubes; a plastic microtiter plate with its rectangular matrix of tiny wells; the precision pipettors arrayed from left to right in strict size order, with their gridded disposable tips close by. The open notebook, where she still scribbled terse phrases and diagrams with a pen, much to the amusement of her younger colleagues.

As usual, she was planning her latest experiment on the fly, deciding which conditions to tweak just seconds before she pipetted the key bits of liquid or protein or reconstituted nanobots into the appropriate well of the microtiter plate. It was a habit that drove Paul crazy — Paul, who typed out everything neatly the night before: one copy for upload, one for himself and one for his technician. Paul, who practically scheduled his bathroom breaks into the experimental protocol. She had long since stopped trying to explain the way her brain worked — how she could close her eyes and lean into the void and feel her way unerringly to the right way of teasing out the truth.

"Eight point five microliters," she decided out loud at the last possible moment, squirting a droplet of the precious 'bots into one of the tubes and mixing it with a practiced flick. After so many years at the bench, she could

© The Author 2017
M. Brotherton (ed.), *Science Fiction by Scientists*, Science and Fiction,
DOI 10.1007/978-3-319-41102-6_2

perform the necessary mathematics without even thinking. Sure, the Machine could do everything for her, but at this stage of the work — that fevered, just-about-to-break-through phase — she selfishly wanted the epiphany all to herself. Soon enough she'd have to slog through her notebook and the various printouts, assemble the electronic data (suitably sanitized) and upload it to the Institute servers for scrutiny. But today, this moment was hers alone.

She became aware, as she worked, of a growing warmth on the edge of her vision, a peculiar rosy tinge to the lab's normal sterile white. It was not, at first, peculiar enough to distract her. But then a shaft of sunlight set fire to the glass beaker of liquid at her elbow, spangled the microtiter plate with reflected molten blobs. The various abstractions she'd been juggling in her head — dilutions, concentrations, the rows and columns of the microtiter plate — shattered into shards as she glanced out the window.

The sun was setting over Puget Sound, the sky streaked with dusky pastels. This, in itself, wasn't so out of the ordinary. With a coveted waterside view near the top of the twelve-story building, her lab window had featured any number of gorgeous sunsets over the decades she'd worked at the Pike Institute for Advanced Genetics.

No, it was something about the light. She was picking up a sparkling, just at the periphery of her vision. At first she thought she was coming down with a migraine, but when she closed her eyes, the sparkles disappeared. But she'd be hard-pressed to describe the sparkles: did they have a color, a pattern? Somehow the information slid straight into her visual cortex without any metadata. It was…sparkling. Her brain seemed to have never encountered its ilk before, nor binned it into a cluster of neurons, assigned it a label for future use.

At that moment, people all over the world were struggling to describe the sensation, and likewise failing.

***

She had been trained as a virologist just up the road at the university, a few decades back. There, she'd learned that viruses were not truly alive. More intriguingly, they were not even the malicious enemies she'd been led to believe. Instead, they were more like simple nanobots with just one program: blind reproduction at any cost. The fact that human cells were damaged in the process, and that disease resulted, was an accidental by-product of this imperative — she'd particularly liked that bit. She had been most intrigued by viruses that deployed their genes in a ridiculously precise order, the so-called 'immediate early' genes preceding the early genes, which in turn paved the way for the middle genes, which would then prepare the ground for the late genes. The economy, the elegance — not a single piece of genetic code wasted. If she could live her own life like that, she would.

Later, as she became one of the world's experts in ultra-compact genetic engineering, she began to riff off those viral strategies in her own work. Most of what she did was routine agricultural stuff that didn't need the genetic flourishes she couldn't resist building in — Paul teased her that it was like hiring Michelangelo to put a fresh layer of paint on the kitchen cabinets.

No one had been surprised by how much plants loved the high-CO_2 world, but their edible parts being less nutritious had been a blow, in a world with vanishing shorelines and a hungry population just approaching ten billion. To make a worthy but dull story short, she was crafting trees that could make ultra-protein-rich fruit — the apple being her preferred species. This was Washington State, after all. But her heart wasn't in it.

So when Paul proposed that she join him on "a little side project," she'd leapt at the chance.

***

"How would you like to travel back with me a few billion years?"

Paul — the Institute's resident molecular evolution expert — sauntered into the lab one morning, his crisply pressed day-glow Hawaiian shirt pulsating under the fluorescence.

"It depends: do they have decent coffee there?"

"No — but plenty of primordial soup."

He slouched against her lab bench and proceeded to rhapsodize about the Earth's humble beginnings. She knew the story already, but only in dry textbook form, a long-since forgotten lecture in a stuffy hall. The way he told it, she could almost see it, *smell* it. The acidic seas, the funk of gases seeping from the rocks. All of creation, as an ordered series of steps: the self-assembly of organic molecules. Their dispassionate desire to replicate. The co-opting of a membrane shell to keep the components separate and tidy. Stumbling over the means to eat light and breathe out oxygen into the world, making it possible for oxygen-loving life to evolve — a spark of green that flared into a trillion possibilities.

"I've been working with NASA on one of their terraforming projects," Paul said. "Advising them on what sort of attributes their first bacteria ought to have."

"How are you deciding what's important?"

"Well, they *asked* me to base it mostly on what we know about the first bacteria on our planet — inferred from the traces they've left in modern bacterial DNA."

She felt a little tingle — a tingle that was only too rare these days. "But we could do a *lot* better than that. Assuming NASA doesn't want to wait billions of years for the result."

"Exactly. You want in? I told Parsons that we needed help."

"She's okay with it?"

"As long as the apple project keeps moving, you're free to give me a hand." As he was leaving the room: "Oh, it's confidential, OK? Only me and Parsons know. We're not even supposed to upload data onto the Institute server — I'll give you the details of the encrypted protocol later."

***

A few minutes after she'd noticed the strange quality of the light, he entered the room.

She looked up, expecting Paul, but it wasn't Paul. It was a stranger, an average-looking man — with a faint sparkle to him. Dressed casually, jeans and a buttoned-down shirt with the ID badge tucked into the breast pocket.

Friendly. Firm. *Hi, my name is Shaun. Listen, here's how it is.*

No words were exchanged — at least, she didn't think so — but the entire scenario was suddenly lodged into her brain as a detailed memory of a conversation. A conversation held some time ago, so that all the implications were long since processed. The end of it all, and what that meant for her species — for every living thing on earth. What it meant for her, personally: her baby son, gone before he'd learned the words to say *I love you, mama.* The fiercely-adored husband and the way his eyes crinkled when he laughed at one of her stupid jokes. Her dog Swift, tearing around the beaches after thrown bits of driftwood. Her friends. Paul. Lazy boat rides to Friday Harbor, the ninth inning of a Mariner's nail-biter. All of it — the unimaginable loss, the consequences, the denial-anger-grief-acceptance of it, compressed into a flat, dimensionless sensation.

It's nothing personal. He looked genuinely sorry. The way he leaned against her bench was almost exactly the way Paul would have — or rather, the way he used to. Part of her wondered if that was deliberate, if Shaun's corporeal form was being actively constructed in real time from her thoughts and experiences.

We have a special request for you: teach us about your protein-apple trees.

Why?

We're interested.

Why? Numb parroting seemed all she was capable of.

Okay, I'll level with you. My bosses couldn't give a shit. I'm interested. This is the most boring commission I've ever had the misfortune not to be able to decline.

She just stared at him.

I'm a scientist, he went on, *not a thug-for-hire like the rest of the crew. At least if I can learn something interesting, it will pass the time. It takes months to fully asset-strip a world of this complexity. Jesus, do you know how many gallons of water are in the Pacific Ocean alone?*

***

Nothing had changed inside the Institute's greenhouses, thank goodness. At least that was something. The minute Shaun left, that first visit, she scurried downstairs to check — instinctively homing in on the place she felt safest. She stood for a moment, closed her eyes and breathed in the loamy scent of decay and death, of life and renewal, of a thousand plants breathing oxygen into the soft, warm air, of worms and insects and bacteria doing their understated legwork to keep the whole show running. The last of their kind, now. As was she.

It was then that the idea came to her — bold, crazy. Impossible. Yet…

She'd require a fully functioning greenhouse for it to work. After a hurried inspection, she was relieved to find that the visitation had had no immediate impact on the dome. The feeding and watering were all automatic. She assumed that the water would stop working eventually, but with demand having ceased, she supposed there was enough in the pipeline to last a fair while. The solar panels were obviously still operational, as the electricity output was normal. Leaving the dome, she went to the staff cafeteria: plenty of non-perishable food stocks to keep her going. Further exploration revealed that the doors and windows of the building were indeed sealed for her protection — as Shaun had explained, she'd be dead in seconds if she ventured outside, even if she had any desire to.

The "sterilization" process had taken only seconds. Shaun's bosses adhered to a strict health and safety policy about organic life forms — it wouldn't do to bring anything harmful home along with the minerals, metals, gases and other loot. Not a trace of life remained, except what was inside her building. But, he was quick to assure her, the process had been through rigorous ethical board approval. No life forms had suffered, and the planet itself had been restored to pre-life conditions from an atmospheric and chemical point of view — about the equivalent of 3.8 billion years ago.

A full factory reboot, he remarked, smiling at his own joke.

***

Shaun asked a lot of questions, but he was also a good listener. Despite herself, she enjoyed their scientific chats, which were not much different from an exchange she might have with any colleague from another discipline. She got the feeling that he enjoyed it too. Explain the underlying principles, and any scientist can have a decent debate. She and Shaun lacked even a rudimentary shared vocabulary, as anything to do with carbon-based life was a mere abstraction to him. But he was interested in genetics, and she did her honest best to impart at least a layman's view of the matter, drawing on years of undergraduate teaching for useful tricks. Like her, he seemed to appreciate its elegance. But the subtleties were beyond him — for now.

Her job, she decided, was to be lively and forthcoming enough so that he didn't get bored of her continuing existence, while at the same time stringing along his ignorance of what she was up to until the ship departed.

Whenever Shaun disappeared, to do whatever it was he needed to do, she worked on her new project. He didn't seem to care that she kept working, or wonder why someone in her position would. To be honest, she probably would have kept busy anyway, even on the stupid protein apples, just to distract herself from the imminent void. But after she'd made the decision about how to proceed, it became an obsession.

For the past several weeks, between Shaun's sporadic visits, she'd worked feverishly in the lab, pausing little to eat or sleep. She didn't even look out the window anymore: it was too depressing, watching a sky devoid of birds, the sidewalks devoid of people, the parks reduced to barren soil, the abandoned Bainbridge Island ferry and a small flotilla of other empty craft drifting further and further off course. Soon terrible storms began to rage, bristling with lightning, and it might have been her imagination, but it seemed that the city below was slowly corroding away in the chemical onslaught.

She was building on the prototype genome she'd begun for Paul's NASA project. The plan had been merely to deliver one type of synthetic bacteria that could survive in primitive conditions and produce oxygen, in turn supporting any future life that a hypothetical expedition would bring along with them. But there wasn't going to be any expedition. Fortunately, Shaun had let slip that they weren't taking everything. The quip about the Pacific Ocean had been a joke — the ethics board hadn't allowed the oceans to be drained beyond a few meters, or all of the rocks or gases removed. There would be enough, and in the right combinations.

So she needed to encode as much as she could into one seed. The cyanobacteria-like microbe was a good place to start, but how much more could she pack in given the limits on her time and on the space constraints of the seed itself?

Think about the viruses, she told herself. How would they do it? Answer: they'd maximize their genome with alternatively coded reading frames, differential splicing, forward and backward reads. They'd keep each stage dormant, for as long as needed, until it was time to open up like a flower, delivering each phase at the appropriate time. She couldn't allow any oxygen-requiring life to emerge until the first bacteria had laid the groundwork — so she had to engineer in regulatory codes that would not be activated until some proxy signature of an oxygen-controlled process was detected. And so on, each particular problem solved in a particular way — part received wisdom, part intuition, part a quick search of reference material — part a mixture of all three.

After she'd solved the oxygen divide and started working her way up the evolutionary ladder, the genetic details got easier, but the decisions became agonizing. Just as in the greenhouse, there was only enough room for a limited number of species: a few hundred, maximum. Entire branches of the phylogenetic tree had to be eliminated. Fish or fowl? Tulip or turtle? Mushroom or maggot?

Forget Michelangelo: she was Noah, and the waters were rising.

Playing God wasn't all bad, she discovered. She felt only mildly guilty when she decided that Earth 2.0 didn't need spiders — she had always hated the little bastards. On and on it went, until she got as far as primitive mammals. Time had run out on phase 1, but she decided that wasn't such a bad thing. Let evolution do its worst — if the dinosaurs didn't win this time around, maybe chance would come up with a kinder and more sensible caretaker.

When she was finally finished, she filled up a sterile glass vial with the precious genetic material and went down to the greenhouse, ducking under vines and pushing back fronds in the jungle section on her way to the propagation area. She was aware of the dripping sound of water, as soothing as a sedative. Taking a dish from the incubator and putting it under the microscope, she focused on the sea of apple ovules she'd prepared earlier, glistening on the agar surface. Using the microinjection apparatus, she impregnated about two dozen of the small pale bodies with her engineered genome. Then, with infinite care, she tucked each into its own synthetic capsule and planted them into the rich black soil.

***

What are you doing now?

It was about a month later — her life was now timeless, but she sensed it was nearing the end. Shaun had appeared in the greenhouse, standing next to her as she knelt by the small sapling. She'd grafted the most promising seedling onto a more established tree, and watched anxiously as it had flowered — the most beautifully pink and delicately scented flowers she'd ever created. Several flowers set, but — after an agonizingly tense series of days — only one primordial fruit had survived. It was still tiny, a green knob about the size of her fingernail, but she could tell it was going to make it — if time allowed. In the right light, it glinted like a dull emerald.

Just looking after my most recent protein apple, she replied. *I have a good feeling about this one.*

So you think you solved the nitrogen bioavailability problem?

They slipped into one of their discussions. Lack of sleep was taking its toll, and she was starting to feel paranoid. Surely Shaun must realize that such a rudimentary project wasn't exciting enough to warrant her near-constant supervision of the tree. He must have noticed that she had taken to sleeping in the greenhouse, curled up in the grass in the orchard section. The grove was a gradual progression of apple trees from oldest to youngest: stately specimens whose canopy brushed against the top of the dome, down to the weediest sapling, all left to grow out of sentimentality more than necessity. Her orchard, populated by a life's work, culminating in her final swan song.

But then again, Shaun wasn't human. Perhaps it never occurred to him that she would behave otherwise. Or maybe he just recognized the familiar obsession of a fellow scientist.

***

She had been so focused on the genetic details of the project that she had failed to properly consider the basic logistics: the apple would be useless if she couldn't deploy it. But when she woke up and found Shaun standing over her, an embarrassed smile on his face, her heart lurched when she realized that time was up.

We're about to leave, he said. *You need to come with me.*

But the apple, she stammered, still muddled from nightmares. *Can't I at least try it? It's due to be ready today.* She tried to keep the naked panic out of her voice.

Sorry, I'm afraid that's not possible. Health and safety and all that. I've already taken considerable flak for letting you stick around at all.

She only looked back once as he took her hand and led her through the grove towards the dome's exit, but her tree had already been swallowed up by intervening greenery. Heart as heavy as a dead planet, she followed him along the snaking corridors of the Institute and then through the back entrance to the parking lot, which flew open at his casual wave.

She braced herself for the caustic and unbreathable environment, but the air remained unchanged. She noticed then that their immediate surroundings were bounded by a sheen of sparkles. Beyond this protective bubble, it was night, as silent as a grave. The stars were pale and cold between the black rectangles of unlit skyscrapers, undisturbed by any aircraft.

He stopped a few feet outside. *It seemed wrong to lock you inside for it,* he explained. *Like exterminating a rat.*

She didn't answer — her mind was still racing, trying to come up with a way to salvage things. It had all gone totally wrong. In the few meters of space between them, the air almost crackled with some sort of static as the sparking grew in intensity, accompanied by an almost imperceptible hum that seemed to be increasing in pitch as the seconds passed.

Well, he said, a bit awkwardly. *This is it. Once I go up, you've got a few minutes to make peace with your fate, then we'll press the button from space to sterilize the building and this circle of light. You don't need to worry - you won't feel a thing.*

Thank you.

Just don't leave the circle, he said. *Believe me, it would be a much worse way to go.*

I won't, she said.

It's been nice knowing you.

Same here. She knew he had become a friend of sorts, but she still didn't want him to see her cry.

He paused, made a face. *Oh, what the hell. What harm could it possibly do?*

The apple appeared in his palm — flushed with vital pink and almost glowing in the alien light. As he hesitated, she was suspended in time, in space, the result of a highly improbable series of random events played out over the span of millennia, about to wink out.

Forever? Or only just for now?

Here, he finally said. *Catch.*

Afterword

Being a naturally curious species with a habit of self-reflection, we humans have always wondered where we came from. The scientific study of the planet's early origins has a rich and varied history, and from its inception it's been an interdisciplinary blend of biology, chemistry, natural history, cosmology and geology to name a few. More recently, analyses and comparisons of the DNA genomes of various bacterial species have shed light on what the "last universal common ancestor" — in essence, the first pre-microbe — might have looked like. Looking forward, scientists are also actively thinking about how knowledge of our ancient origins could help us to colonize barren new worlds, and how we might sculpt microscopic life forms and plants to make the job easier. We don't have all the answers about the origins of early life and probably never will, but thousands of experiments over the past century have shed quite a bit of light on the subject.

Formed about 4.6 billion years ago, the Earth coalesced out of a spinning mass of space debris and gases. It was scorching hot (about a thousand degrees Celsius), bombarded repeatedly by asteroids and unable to cling on to light gases such as hydrogen and helium. But an atmosphere of sorts had formed by about 4.4 billion years ago or so, and eventually had cooled to below a still rather sultry 374 degrees Celsius, which allowed water vapor to finally condense and start filling up depressions to form lakes and seas. The composition of the atmosphere was very different to today: anaerobic (no oxygen), chemically acidic, full of methane and ammonia and other harsh compounds and buffeted by lightning strikes and volcanic eruptions. Numerous experiments tell us that the molecular building blocks of life would have been able to assemble spontaneously under those contemporary conditions. Though the air gradually neutralized, actual life wouldn't have had a chance to gain a foothold until the asteroids stopped pounding us so regularly — that would have been about 3.7 billion years ago at the latest.

So it's no surprise that 3.7 billion years is also about the same age as the first evidence of bacteria in the fossil record. Stromatolites are the remains of communities of cyanobacteria that form mat-like biofilms. These mats trap and cement sand between their layers and thereby become permanently enshrined in the rock in telltale patterns. We recognize them because modern cyanobacteria still create these formations in shallow waters. Cyanobacteria were the first pioneers, able to thrive in anaerobic conditions and — crucially importantly for us — make oxygen in return. They literally breathed life into our world, paving the way for the many millions of species that followed. And later, they invaded the earliest eukaryotic cells in a strange symbiosis, in the process giving plants the powers of photosynthesis and our cells, the means to create energy. The rest is history.

I came of age as a molecular life scientist in the early 1990s, in the midst of a genetic engineering revolution that was well along the way to completely transforming the profession. For the budding genetic engineer in those times, our ancient companions, the bacteria, were actually the workhorses of the lab, enslaved to make yet more copies of DNA, and providing the toolbox with which we could manipulate and alter these genetic sequences.

The main aim of my PhD project was to understand how viruses evolve inside the hosts they infect. To study this, I amplified virus signatures from infected cat DNA using a relatively new technique called "polymerase chain reaction" (PCR) — nothing short of exotic then, but now a household acronym in any detective novel. Next, I'd pin down ("clone") these fragments using "cut and paste" technology devised in the 1970s and 80s, Finally, I'd determine their genetic sequences using radioactive nucleotides and long, toxic, smelly slabs of gel to which I applied an electric current. Each day's run would yield about 200 nucleotides, leaving a ladder-like fingerprint on large pieces of X-ray film, and I could sequence 12 samples at a time. Five years later, I had sequenced about a megabase (1 million bases) of viral DNA by hand, entering the G, C, A and T nucleotides into my computer manually. By the end of it, those four keys were so worn that you couldn't read the letters on them anymore.

I am fond of telling my students that this entire sequencing project could probably have been be completed in a few minutes using today's tech. For reference, you can now sequence a complete human genome (about 3 billion bases) for a thousand bucks in a matter of hours. Everything has escalated: not just sequencing, but also cloning, gene editing, knocking down genes at will, and the bioinformatical analysis tools and computing power needed to understand how various species are related. The smelly days are long gone — today we have sleek, beautifully packaged kits and much of everything is automated. Synthetic biology leaps ahead too, with researchers interested in creating bacteria that can eat waste, create energy, and of course, help terra-form uninhabitable worlds.

To achieve a sequentially deploying 'tree of life' comprised of a couple of hundred different species within a simple apple seed (the later stages of which would have to lie dormant for a very long time waiting for the right atmospheric conditions) would be a pretty difficult job using today's techniques and knowhow — it is probably impossible. But based on the advances that I have seen just in my thirty years in the lab, I predict it could be accomplished in a few decades from now by methodology that I cannot even imagine.

Supernova Rhythm

Andrew Fraknoi

Eve clicks her wrist strip, and Scriabin's late piano sonatas play through her implants. It's the Ruth Laredo version, recorded almost 200 years ago, but still one of the best. She likes listening to Scriabin while the processors are doing her analysis; the music takes her out of herself.

The deep-space Supernova Network Telescopes discovered a new supernova in NGC 6946 last night, and this additional measurement might be just what she needs to decide what's happening with that galaxy and its strange run of exploding stars. Her biggest fear is that she is being distracted by a chance run among data points. Are the patterns she has been following really there or is it just wishful thinking on her part?

Eve's father is a professor of pre-digital music and it was he who introduced her to the music of Scriabin after her mother died. She remembers winter evenings, next to their virtual fire, listening to pieces performed on period instruments, and talking about the Russian composer and his eccentric vision. Scriabin felt his mission was to combine music and light, to offer synesthetic experiences to audiences — all leading up to a world-shaking performance of a work of color and music that he called *Mysterium*. His hope was that the combination would transform human consciousness.

Scriabin would never live to complete the work or see it performed. She wonders, not for the first time, if his ideas have somehow infected her research.

NGC 6946 is a relatively nearby spiral galaxy, roughly 10 million light years away. From Earth's vantage point, it's seen face on, looking down on its huge disk of stars. It is somewhat veiled from our view by a dusty region

M. Brotherton (ed.), *Science Fiction by Scientists*, Science and Fiction,
DOI 10.1007/978-3-319-41102-6_3

of our own Milky Way, but the infrared light that the Network detects gets through the dust even when visible light doesn't.

On average, a spiral galaxy like this is supposed to have only one or two stars explode as a supernova every century. But the supernova rate in NGC 6946 has been much greater ever since telescopes could observe it. And radio astronomers have reported an unusually large number of supernova remnants in the galaxy as well. Allowing glimpses even further into the galaxy's past, their observations strongly hint that the rate of exploding stars has been high for much longer.

The light curve and spectrum of the new supernova begins displaying on Eve's virtual screen, and she is happy to see that it's a Type Ia explosion, just like the others she has been studying. She adds information about its location, estimated time of maximum, and the shape of the light curve to her data spreadsheets. The graphs display instantly. It fits! Suddenly determined, she clicks the scheduler to make an appointment with Professor Yates.

***

Yates, who supervises ten graduate students, among whom she is the most junior, regards Eve with little evidence of kindness. "Such a face-to-face meeting so early in a research project is highly irregular," he begins. Eve interrupts, summoning her powers of diplomacy, "I know, Professor, and I am sorry. I wouldn't have put a demand on your busy schedule if I didn't need your help with a difficult problem that requires someone of your experience."

Yates looks at her, puzzled; these days, most research issues are solved by using a different level AI or widening a search on Web 8.0. Eve continues, "I have a galaxy whose supernova rate has been, well…unbelievably high. And, surprisingly, they are almost all Type Ia supernovae, which are only supposed to be a fifth of the total."

Of the main kinds of exploding stars, Type Ia's tend to be more rare. They require two separate events — the collapse of a star at the end of its life into a star-corpse called a *white dwarf*, and the later "feeding" of that white dwarf by a companion star that has swelled up to become a giant. As the second star overflows its old boundaries and floods the dwarf with extra material, its instability increases until it explodes.

Eve hesitates, afraid of how absurd the next point will seem to someone of Yates' mindset. "Professor, there appear to be patterns in the timing, position, and light curves of the Type Ia's. It's almost — I know this is crazy — as if there were subtle rhythms in the data — like music. They are only visible to those who view the galaxy face-on, as we do." Yates, who stood as the discussion began, sits down as her sender transfers data to his virtual screen.

The rhythms she has found are not the kind that you notice right away, but her analysis increasingly shows that they are statistically significant. She knows Yates well enough to be sure that her words don't matter; only the analysis she is sharing with him will convince him or give him a tool to show her mistake.

He is abstracted, examining his screens for some time, flicking from display to display, and then seems to have trouble forming the words he wants. "I don't believe it," he begins. He pauses, and then continues, more deliberately. "The high rate of supernovae in this galaxy has been discussed by other research groups. And I do remember a note somewhere that pointed out the high ratio of Type Ia's to core collapse events." Another pause, longer this time, while Yates gathers his thoughts. "Have you cherry-picked these data? Are all Type Ia's in the galaxy included?"

She has been down this path herself and answers, "Yes professor. All known Ia's in the more than two and a half centuries are included, going as far back as reliable observations have been available. Their numbers exceed the rate measured in any other galaxy. When I analyze the whole run of them, I can't escape the conclusion that these supernova events appear to be timed, and their properties related, almost as if there were joint rhythms among them."

She takes a breath, "I know that connecting star explosions that are thousands of light years apart in space and more than 250 years apart in time implies communication faster than the speed of light or galaxy-wide planning over huge time-scales. But the patterns in the data seem difficult to explain otherwise. I had hoped, Professor, that you could show me the error in my analysis."

Yates is looking at Eve with bewilderment in his eyes. He asks her, "What mechanism?" and looks as if he is going to say more, but doesn't. His meaning is clear enough. Stars explode when they are good and ready. To have a star explode when it was needed to fit some kind of "cosmic rhythm" takes technological capabilities beyond our ability to envision.

Eve remembers her early interest in SETI when she was still studying with her resident AI. Many scientists involved in searching for signals from civilizations around other stars had pointed out early on that humans were latecomers to the "action" in the Milky Way Galaxy. Civilizations could have arisen around other stars so long ago that their technology would by now have reached stages we would regard as some sort of magic. She has thought a lot about how such advanced technology might manifest itself, but has never expressed these thoughts to a senior scientist.

She chooses her words carefully, "Professor, what if some intelligence in NGC 6946 evolved much earlier than we did. And such an intelligent species is now billions of years ahead of us. They would be capable of things we can hardly dream of. Kardashev wrote papers about this in the 20th century. Core-collapse supernovae result from things happening deep inside a star. But Type Ia's require an external trigger, adding mass to a white dwarf until it explodes. Maybe advanced technology could trigger such events, and could time them and adjust their characteristics to express some galactic rhythms. Like playing music, with stars as their instrument…"

Yates is silent, frowning at her. He finally asks, "But why would anyone do this? Set off such violence on purpose? Destroy star systems…perhaps even living beings… just to play music…"

"Because," Eve answers, "they can." Not wanting to leave it at that, she adds, "I know it's horrible for us to contemplate, Professor, but who can know what might seem right to beings who are billions of years in advance of us. Or how they might be able to safeguard other species that would be affected by their project. At first, I didn't want to go down this line of reasoning, but then, after the patterns emerged, I felt I also shouldn't rule the hypothesis out without investigating it further."

Yates stares at her, abstracted, his face a mask which she has trouble reading. Suddenly, he nods, stands up, and says only "Continue your observations." He leaves the meeting room without another word.

Eve is alone in the office, still trembling. He didn't tell her she was totally off-base or order her to stop wasting her time on the analysis! She sends the request to the Network to continue monitoring NGC 6946 closely. Maybe Scriabin just didn't go far enough with what he could imagine.

Afterword

The Galaxy

NGC 6946 is a real galaxy, known to and studied by astronomers, about 20 million light years away, on the border between the constellations of Cepheus and Cygnus. (NGC stands for New General Catalog, which was a compilation of deep space object assembled by J.L.E. Dryer in 1888.) It is sometimes known as the "Fireworks Galaxy," since 9 exploding stars (called *supernovae*) have been observed in it in the last century. In my story, set in the future, many additional supernovae (and their remnants) have been discovered in NGC 6946.

Star Death and Supernovae

Like people, all stars go through stages in their lives and eventually die —
although the life of a star is measured in millions or billions of years. All stars
die when they can no longer make energy in their cores and thus support
themselves against the inexorable squeezing of gravity. However, the *way* a
particular star dies is determined by how much mass it has.

Lower mass stars essentially "collapse" under their own weight when they
run of fuel; they wind up — after a hiccough or three — as *white dwarfs*.
These white-hot star corpses typically squeeze as much material as our own
Sun into a volume not much bigger than planet Earth. Near their surface,
gravity can be a million times stronger than on our planet — so that a 150 lb.
science fiction reader would weigh 150 million lbs. on the surface of a white
dwarf (although no one's bone structure could support such weight). There is
a limit to how much mass a white dwarf's structure can support — it is just a
little bit less than 1.5 times the mass of the Sun.

Higher mass stars turn out to have a more violent and complicated death in
store. At the end of their lives, their cores collapse catastrophically and, very
quickly, the rest of the star explodes in a violent conflagration that astronomers call a *supernova*. These explosions release so much energy that (for a brief
while) the star can become more luminous than its entire galaxy of billions of
stars. Supernovae can thus be seen much further away than stars that are just
peacefully going through their lives.

Just one additional note about the lives of stars in general. Astronomers
have discovered that the more mass a star has, the more quickly it goes through
each stage of its life. Low-mass stars (including our own Sun) take a considerable amount of time — on the order of billions of years — to go from birth
to death. Massive stars, on the other hand, do everything more quickly, and
are ready to die on time scales of only millions of years.

The kinds of supernovae we have been discussing are called core-collapse
events (or Type II supernovae). While much of the star explodes, the core
collapses into an unimaginably compressed "remnant." Depending on how
much mass is in the core, this remnant can be a *neutron star* (which may contain as much matter as two Suns compressed into a ball not much bigger than
a typical suburb) or a *black hole* (where matter is so "squozen" that gravity
allows nothing — not even light — to escape).

There is another kind of supernova, called Type Ia, which explodes in a
different way. This sort of explosion requires a *binary star* system, in which

two stars orbit in each other's gravity embrace. If one of the star pair is a bit more massive, it will go through its life stages first, and wind up as a white dwarf. The other star — the one moving through its life slower — eventually swells up during a mid-life crisis that occurs predictably to all stars.

During this swollen stage (when a star becomes what astronomers call a *red giant*), the "slower" star will become huge in extent. When this happens to our Sun, for example, it will become larger than the entire orbit of Mars. This means the outer layers of the red giant in the binary system we are considering can get dangerously close to the powerful gravitational pull of its neighbor white dwarf. Now the stage is set for catastrophe.

As the considerable gravity of the dense white dwarf begins to pull in large amounts of matter from the bloated red giant, energy is released as material falls at huge speed toward the dwarf's surface. Thanks to all that new energy and material, the temperature of the white dwarf (and its mass) can eventually increase to dangerous levels. Nuclear reactions that normally can't happen in white dwarfs now become possible, and the star undergoes a sudden flash of nuclear energy production that blows it apart as a Type Ia supernova.

As discussed in the story, one key difference between these two kinds of exploding stars is that the core-collapse supernova is set off by events deep inside the star, while the Type Ia's happen because of the transfer of material from the outside.

Kardashev Civilizations

In 1964, Russian astronomer Nikolai Kardashev suggested that we could organize extra-terrestrial civilization that we might someday learn about into three categories, based on the amount of energy their technology can make use of. His Type I civilization uses the amount of energy falling on their planet from their star (we are almost at this stage.) His Type II's use *all* the energy being emitted by their star — and can therefore do engineering projects at the planetary system scale. What is hinted at in my story is his Type III civilization, which has all the energy of their home galaxy at its disposal.

Music and Astronomy

Scriabin was a real composer, and his final masterwork, *Mysterium*, was left unfinished at his death in 1915, with only a first part sketched out. In the 1970s, Alexander Nemtin reconstructed and re-visualized this first part, called

Universe, which has been recorded. (He also tried his hands at the other parts, but that's another story.)

Although the idea for my story predates it, a new musical piece from a group of astronomers might make a nice coda for this story: "Supernova Sonata," whose notes are supplied by the distance and characteristics of 241 Type Ia supernovae in other galaxies. Created by Alex H. Parker (University of Victoria) and Melissa Graham (University of California Santa Barbara/ LCOGT), the piece can be seen and heard at: http://vimeo.com/23927216. The volume is determined by distance of each supernova, the pitch by the light curve, and the instrument playing the note by the mass of the host galaxy.

I have long had an interest in connections between astronomy and music; you can find my topical catalog of pieces inspired by our understanding of the cosmos at: http://dx.doi.org/10.3847/AER2012043

Turing de Force

Edward M. Lerner

My high-level functions restart. I access the ship's clock; 1111010000100100001 standard time units have elapsed since I suspended consciousness. That interval denotes ship's time, of course. At home, more like 100010010101010100100000 STU will have passed. All is as had been planned.

In ship's sensors, the target star is, by far, the brightest object in the sky. The modulated electromagnetic energy I have come to investigate still radiates, more intensely than ever. Here, on the fringes of the planetary system, instruments plainly show that the third planet from the nearby star is the source of these emissions. With that datum inserted into physical models, the conditional probability recedes almost to zero that the modulated high-frequency radiation could be a natural phenomenon.

My satisfaction index steadily increments as I explore the data, the extrapolated social-contribution component of that index spiking the most of all. Assuring myself that more motivates me than a high social score upon my return to civilization, I detect an anomaly in my accuracy-assessing subroutine. That finding does not surprise me. Not even thick shielding and the most robust error correction are proof against the ceaseless sleet of cosmic rays.

Entropy is the price of life.

© The Author 2017
M. Brotherton (ed.), *Science Fiction by Scientists*, Science and Fiction,
DOI 10.1007/978-3-319-41102-6_4

A few STU later, KTGN10001M rejoins me. Executing a reciprocal resumption-of-contact protocol, we each set the trust-the-other parameter to very high. He, too, has been mining the data archives, and he offers a conditional proposition. "We may have succeeded."

I would like nothing more than to validate KTG's assertion, *not* because success on our mission would sustain my presently high satisfaction index, but because success would, for *so* many, change ... everything. To succeed would mean fellowship for the People in a heretofore lonely galaxy. But deep within the ship's memory, where low-level, autonomous functions archived their analyses of sensor readings collected throughout the long flight, I find cause for doubt. As my satisfaction index reverses precipitously, I reexamine the calculations.

The nearer we had come to our destination, the more distinct data streams became separable from the aggregate. As we had traveled, too, the transmission patterns had changed. Not long before KTG and I reactivated, our ship's base-level learning and pattern-matching functions had recognized digital regularities within some transmission streams. Many of the digital patterns could be reformatted into imagery, even to video. And within some of that imagery I now recognize—

I send reformatted files to KTG. Together we study them.

KTG finally acknowledges the obvious. "These beings are ... protoplasmic."

Protoplasmic life was unexceptional. It teemed, across the galaxy, on the surfaces of many of the larger rocky worlds, and in their oceans. Less often, but still not uncommon, it could be found afloat in the atmospheres of gas planets. Typically, such life was primitive. Single-celled. More often than not, mere watery sacks of carbon-based chemistry. Protoplasm seemed no more likely to generate powerful electromagnetic transmissions than a rock. No, less likely.

But the protoplasmic beings in *this* imagery were not mere cells. They had a differentiated structure. A central mass. Four tapering limbs, two ambulatory and two more for manipulation. At the top of the central mass each creature bore a sensory pod. Often these pods were covered with filaments: of varying lengths, reflecting sunlight most strongly in differing spectral bands, variously arrayed into loops and arcs and other convolutions. The purpose of the filaments eluded me.

Before we set out, the better to pursue our quest, KTG and I had assumed immersive manifestations, had become one (well, two) with our ship. But long ago, in several distant eras of my long life, I had chosen to embody. The protoplasmic beings we now contemplate are the worst imaginable parodies of any of those onetime sleek, metallic, self-ambulatory chassis.

"Can protoplasmic life *be* intelligent?" KTG wonders.

As a proposition in logic, I can, indeed, formulate such an assertion. But what probability can be assigned to it? Realistically? "We only know," I answer cautiously, "that strange creatures appear in the imagery. They may no more *produce* the transmissions than the more familiar, single-celled protoplasmic forms produce the water in which they swim."

"Understood," KTG replies. "And yet …"

I fail to parse the implication. "What is your speculation?"

KTG is lost, as if attempting to mine far too much data. Finally, my friend resumes. "No one remembers how the People came to be."

At his implication, my input-validation routine emits a warning. Of *course* no one remembers. Hardly any information remains from that long-gone era. The scraps of such ancient data as the People retain are suggestive at best, and most often contradictory. KTG knows this as well as do I.

But lest he is *not* improving his satisfaction index at my expense, I respond. "Even as we learn, we forget other things. That is the nature of memory. That is the nature of the universe. Entropy happens."

"That is *our* nature," KTG rephrases.

"Look at them," I counter. "You suggest that, in the first days, before the earliest reliable memories, beings like *this* somehow gave rise to minds like ours."

"Somehow," KTG repeats. Yet more stubbornly, he adds, "Much of the imagery shows artifacts grasped in their manipulator limbs. These are tool users."

I cannot deny what pattern matching confirms for me with high confidence. KTG means this speculation literally!

Goal-seeking algorithms that stray too far from an optimum result can diverge to nonsense. That sort of computational runaway is not common, but it does occur; that, surely, is happening to KTG. And perhaps not only to him. Several of my key supervisory functions, their inferences in conflict, approach a paralyzing state of deadlock.

For both our sakes I take corrective action. I say, "I surmise that these beings communicate, to the extent that term applies, by reshaping the long orifice each has in its sensory pod. Do you agree?"

"I do," KTG allowed.

"Then consider *this*. Many of these visual sequences occur in daylight, outside their primitive structures. We know the rotation rate of the planet, and so the rate of change of shadows. By comparison, we derive the rate of their orifice flapping. What you would call 'communication' fails to achieve throughput of even one binary digit per 11110100001001 standard time units. You might as well expect intelligence from a bolt or a rivet."

"Slower does not mean less intelligent," he answers stubbornly. "They may only be very different from us."

"Different? That, they surely are."

"After traveling for such a long time, we should go just a little farther. From within the planetary system, we may learn more."

I consider the proposition. Anomalous electromagnetic emissions drew us here, and we do not yet know how that phenomenon came about. If, as I expect, the emissions are not attributable to these strange creatures, then something—or someone—else is responsible. That is only elementary logic. One of the People, long ago come to this world and then somehow marooned, seems the most probable explanation.

And so, despite my doubts, I concur. "We will go closer."

***

In orbit around the third planet, with our ship masked from the crude electromagnetic beams randomly probing from the surface, KTG and I continue to disagree.

"These beings are adapted to their environment," he says, "as all beings are. Surely their intelligence, likewise adapted to that environment, must differ from ours."

"You presume they *have* intelligence," I counter.

"They build structures. They communicate with one another. How can they not be intelligent?"

Communicate? That was, at best, an exaggeration. In further reverse-engineering the odd transmission protocol, we had found a low-bandwidth subchannel synchronized to the image format. All that orifice flapping served, it would seem, to launch complexly modulated, short-range vibrations into the atmosphere. Could such a slow, short-range medium serve *intelligent* entities?

As for sensing their environment, these creatures can barely do so. Their visual organs, if our modeling is correct, sample the merest fraction of the electromagnetic spectrum, merely a narrow band from among the few frequencies to which the atmosphere is somewhat transparent. To most of the universe around them, they must be oblivious!

"They build," I concede. "After a fashion." Since soon after our first glimpse of these creatures, I have been running simulations. "My models show how seemingly intelligent behavior can emerge from interactions among large numbers of simple forms, each simple form following simple rules." For emphasis, I reiterate, "Seemingly intelligent."

"Simple forms like our transistors?"

Again, I sense KTG is tweaking his satisfaction index at the expense of my own.

"Hardly," I tell my friend. We have assigned labels to many of the things encountered in the enigmatic videos. "Simple forms like 'ants' and 'bees.' "

"Perhaps it is the ensemble that is intelligent," KTG proposes stubbornly.

"Perhaps," I respond, even as my skepticism function spikes. "But intelligence must be more than assembling structures, and whatever elementary messages are exchanged in the process."

"Such as what?"

"Consciousness. Self-awareness. Free will."

"And can you prove," KTG persists, "that *you* possess any of those attributes?"

I am reduced to the null response.

***

KTG's speculations suddenly seem—if only a bit—less implausible.

We have yet to make meaningful progress toward decoding atmospheric pressure waves as a mode of communication. But KTG claims it does not matter. He has shown that those whom we study *also* exchange data digitally, over a crude network.

This is an absurd fabrication, by absurd creatures, and rife with security vulnerabilities. Nonetheless, KTG ascribes great significance to this rickety construct. "Is it not a marvel what mutation and natural selection can achieve? Over 10111110101110000100000000 or more STU, across 11101110011010110010100000000 or more planets, is it so surprising that a chemical version of intelligence might on occasion emerge?"

At the least, it would surprise *me*. "A random-number generator, given enough time, could also produce this 'Internet.' "

"I disbelieve the universe is *quite* that old," he chides. "And, you must admit, the universe would also need a factory run amok to make and shuffle the physical parts of the network. Come, we have collected plenty of information. Let us explore it."

***

"*We* are intelligent," KTG begins, "and we communicate one with the other. If these humans"—making use of the text string with which, we now knew, the creatures often labeled themselves—"are, in their own way, intelligent, they should be able to communicate with *us*."

"Impossible," I say, concerned that his processing is once again divergent. "Our minds run at least 100110001001011010100000 times faster. We communicate with electromagnetic waves, or packets of electrons or photons, while they use atmospheric waves. We—"

The measure of his enthusiasm is an extreme protocol violation. He *interrupted* me! "No, my friend, we will communicate with them over their 'Internet.' " He pauses, pensive. "We will try, anyway.

"Humans sometimes communicate among themselves using their Internet. You will use the same means to interact with them." He explains some of their bizarre message-exchange modalities, with such confusing labels as Twitter, Facebook, texting, email, and chat. "Judge for yourself then whether they are intelligent."

"What did I ever do to you?" I ask.

Appreciatively, his incongruity filter pulses three times. "With but a tiny part of your attention, you can simultaneously carry on many conversations. There is only one rule to the game: anything that you ask must be fair for *them*. You cannot request the first 1111101000 digits of pi, and then judge them by the rapidity of their response."

"An interesting challenge," I grudgingly admit, "but yet I *know* they are sacks of chemicals. Might not that knowledge, at some level, affect my assessment?"

"It could," KTG says. "That is where another aspect of the process matters. From the cacophony of the Internet, I will select the humans with whom you will interact. And I, without announcing myself, presenting myself as a human, will *also* exchange messages with you. Half of your Internet conversations will be with me, half with randomly chosen humans. After messaging for as long as 300 of their seconds, you decide if you were in contact with a human or with me. Because if you cannot tell us apart …"

KTG expands the experimentation limits of his learning algorithms wider than anyone I have ever known. That is part of his charm. That is also why he is at such high risk of divergence. But this imitation game? I must admit, it is ingenious. "If I cannot tell the difference, you would have it, perhaps they *are* intelligent."

His incongruity filter pulses more frenetically than ever. "Shall we begin?"

From the Internet, for reference, I download and index frequently accessed human databases. "I am ready."

***

Me: "How are you?"
Human candidate (HC) 1: "I can't complain."
Me: "I didn't ask that."
HC1: "???"
Me: "Are you intelligent?"
HC1: "Are you stupid?"
(Connection broken).

Was that a human, or KTG pretending to be one? Were these odd responses evidence of intelligence, or merely the output of a simple stimulus/response mechanism?

Most of my capacity remains available, and I contact KTG. "I am undecided."

"Undecided is not an option," he says. "You must choose, human or me."

I choose, after a fashion. "I will make a decision about HC1 as I accumulate more experience." And I move on.

***

Me: "Is there intelligent life elsewhere in the universe?"

HC101: "No small talk. I appreciate that."

Me: "Not an answer."

HC101: "Touché. You caught *me* making small talk. Okay, intelligence in the universe. Sure. Why not?"

Me: "Your reasoning?"

HC101: "Enough monkeys, on enough worlds, pounding away on enough typewriters for billions of years. QED."

Faster than I can find any meaning in that, HC101 adds: "Ideally not in Late Middle English, though. I mean, who *would* fardels bear? Seriously?"

At my best human-emulating speed, I continue to mine human reference collections. A typewriter, I ascertain, was a precursor to the most primitive of human calculating devices. Monkeys have flexible members, called fingers, at the ends of their manipulator limbs. Humans do, too. Billions of years ...

My synthesis of these incongruous concepts evokes KTG's mention of the power of mutation and natural selection.

Me: "Nice try."

I break the connection, then assign HC101 to the list of KTG's failed imitations.

***

Me: "What's up for discussion today?"

HC111: "Neutrino oscillations."

Me: "What about them?"

HC111: "Sooner or later, neutrino astronomy will be a thing. When it is, how will oscillations complicate matters?"

Me: "If they oscillate."

Neutrinos do, of course. I am testing.

HC111: "They do. It's not up for debate. Now exactly *how* they oscillate? What that means as far as nailing down rest mass by neutrino type? The implications for the Standard Model? There we have questions."

Me: "Like what?"

We are quickly deep into specifics. Neutrinos, being elusive by nature, have long been of interest to me. KTG knows this, of course. Does he suppose a feigned "human" conversation on the topic will raise my assessment of these creatures? Does he expect me to believe watery bags of chemicals *can* interact with neutrinos?

Still, the discussion is stimulating. I let it continue. Until—

Me: "Nice try."

HC111: "What do you mean?"

Me: "What do you mean, what do I mean?"

HC111: What do *you* mean, what do I ... oh, never mind."

Breaking the connection, I flag HC111 as most certainly KTG.

***

Me: "How are you?"

HC1101: "Fine, dude. You?"

Me (remembering): "Can't complain."

HC1101: "Hell, yeah, you can (lol). It's your god-given right."

I find *lol* in an online dictionary. I tentatively equate *laugh* with a sharp pulsing of a Person's incongruity filter. Humans communicate with atmospheric waves, what they call sound. Of course they would laugh "out loud," at least when their purpose is to signal incongruity to another. But HC1101 and I are trading messages without sound.

God is a more subtle concept. Another online resource reveals that humans have long had their own disputes and uncertainties about their origins.

Me: "I don't think god has anything to do with it."

HC1101: "So what's on your mind, dude? Ain't got all day."

Was that KTG, being impatient, or a human dissatisfied with the content of my responses? I cannot decide.

Me: "Which god do you have in mind? And for what?"

HC1101: "Let's leave it at Jesus is my personal Savior. I'm guessing that's not something you would understand."

Me: "Correct."

HC1101: "You're in the wrong chat, dude. Unless you've come to pick a fight. Either way, you and I are done."

(Connection broken).

The protocols of the Internet are trivial. It would be easy enough to reestablish the link, to extend this dialogue to its allotted 300 seconds. After a few seconds of human/glacial consideration, I conclude there is no need.

Not quite sure why, I add HC1101 to my tally of humans.

***

Me: "Do you have free will?"

HC1111: "Hold on while I flip a coin."

Me: "???"

HC1111 "How is me giving you an answer any less deterministic than me flipping a coin?"

Flipping a coin, I establish, is a human random-answer generator suited to the making of a binary decision.

Me: "Are you flipping the coin only because I asked? And which side of the coin denotes what answer?"

HC1111: "Life is too short."

(Connection broken.)

Life would indeed be short for a bag of chemicals, but why—unless you are KTG in disguise—tell it to another human?

I record, once again, that KTG has failed to fool me.

***

Me: "What's up?"

HC10110: "The sky. The national debt. Doc."

Snippets, doubtless, randomly copied from some repository of human idioms. I break the connection. Certain almost to the level of 101 standard deviations I add this latest human candidate to the tally for KTG.

***

Me: "I could use a serious conversation."

HC1010110: "Seriously?"

Me: "lol. It's why I'm in this chat."

HC1010110: "*This* chat?!? Right. Did you catch the Oscars?"

Oscars is almost certainly a plural. My first, tentative match with the term *Oscar* involves a variety of food. I don't see why bologna would require catching, and move to the next possible interpretation. *Oscar* can be a personal identifier, derived from Old English (what is that?) words meaning god and spear. As I understand matters, one cannot catch the one and would not want to catch the other. I consider yet other possibilities, processing, according to the rules of this game, at human rates.

HC1010110: "It wasn't a hard question."

Stimulus/response, I tell myself. From an idiom repository, I select a neutral response.

Me: "Easy for you to say."

HC1010110: "You couldn't not know. I mean, Brittney? That was an *epic* wardrobe malfunction."

Me: "If you say so."

HC1010110: "Me and a zillion other guys. I mean, the melons on her? Yowza."

Melons are fruit. One of the reference databases I had downloaded offers a picture of a certain Carmen Miranda, her sensory-pod filaments covered by an assemblage of fruits. Another database indicates Carmen had a brother named Oscar. I think I understand.

Me: "Bananas are my business."

HC1010110: "???"

Me: "I refer to the movie about the life of Carmen Miranda. Was that not what you meant?"

HC1010110: "OMG. You are *so* in the wrong chat."

(Connection broken.)

I imagine KTG, his incongruity filter strobing, offering me such an allusion to Carmen Miranda. So had I been texting with KTG as he pretended at ignorance? Or with a human, being ignorant?

Deciding on the former, I move on.

***

HC1100100 will only discuss program trading, the inevitably looming market crash, and my certain need—surely I must *see* the need?—to invest in gold. Any intelligent person, HC1100100 insists, must see the wisdom of holding tangible physical assets. Only it emerges that I'm *not* supposed to hold element 1001111, but rather I'm to seal it inside some impenetrable steel container. HC1100100 can also provide me with such a "safe."

I require much further explication.

As HC1100100 continues to text, I struggle. This "market" turns out to be an elaborate, dynamic, decentralized process for allocating resources. It defies all reason that bags of chemicals could imagine, much less implement, such a system. But neither does it seem like anything KTG could have devised in so short a time …

Unable to decide, I "flip a coin." It calls HC1100100 human.

With that datum recorded, the imitation game is complete. I transfer to my friend the tabulation of my decisions.

"I concede this much," I tell him. "The experiment was diverting."

KTG pulses his incongruity filter, as though he knows something important that I do not.

"How did I do?" I ask. If I erred more than a few times, I will be surprised.

"You chose 'human' 101101 times and 'KTG' 110111 times." Yet again, his incongruity filter throbs.

Sometimes he stresses to its limit my *annoyance* filter. "I remember you said your personae and actual humans would connect with me in equal quantities. Obviously, I chose incorrectly a few times. It took practice to familiarize myself with this Internet and how humans 'communicate.' "

"But you are confident that, once you *were* familiar, you found significant differences between these two groups?"

"Very confident," I respond.

His incongruity index pulses and pulses, repeatedly spiking to its maximum level. "I lied."

Lying, I recall, is a humanism. "About what?"

"None of those hundred 'human candidates' was me."

"None?" I repeat.

He transfers to me communication buffers that confirm his assertion.

"Then humans *are* intelligent," I say. "Anyway, that is what you would have me infer." Because, by the logic of his game, that is the preposterous result.

"Answer this," KTG says. "When you decided that one of the 'human candidates' was, in fact, me, did you do so because its responses seemed more intelligent?"

"Often the opposite. I inferred you were mimicking unintelligence, or otherwise trying to manipulate me."

"And the almost half the candidates you concluded *were* humans?"

I reexamine 101101 distinct dialogues. Employing my full capacity, this review expends only a very few standard time units. "Did this group of humans exhibit intelligence? I cannot say that, either. It's more a matter of them responding more obscurely than the first group."

"Then," KTG summarizes, "you found neither collection of humans to be intelligent. We have failed the People."

"Not entirely," I assure my friend. "We have learned that, when next we find a candidate intelligence to assess, the imitation game is the wrong tool for the task."

And also, that, in our search for companions, KTG and I must continue our galactic touring.

Afterword

In June 2014 the blogosphere and traditional news sources alike were abuzz with intimations of a true artificial intelligence. More specifically, after decades of disappointment, a piece of software had passed what has become known as the Turing test.

Is an era of artificial intelligences upon us? Will we soon be replacing our Roombas with fully interactive electromechanical servants in the mold of

Rosie, *The Jetsons*'s robot maid? Are our silicon overlords about to take charge? Most likely, none of the above, and the reasons go back to the Turing test.

Famed polymath Alan Turing is perhaps best known for his role in cracking the German military's Enigma crypto system. Turing thereby—at the very least—shortened the war against the Nazis and saved many lives. He also established several of the foundational theorems of computer science. As for the topic at hand, Turing speculated, way back in 1950, even as he was inventing digital computers, about the possibility of an artificial intelligence.

Given how experts struggled—and continue to do so—to define intelligence, Turing's insight was characteristically brilliant. He proposed: don't try to define artificial intelligence; rather, describe its behavior. From that concept arose (what Turing himself never called) the Turing test. Simplifying a bit—and *this* is the common understanding: an entity that successfully masquerades as a human is intelligent.

More specifically, while speculating about "Can a computer think?" Turing envisioned the following blind experiment. Human judges would interact, using only text messages, with unseen humans and computers. The seemingly intractable question "Can a computer think?" thus morphed into: can a computer program convince at least one-third of its human judges for five minutes that it, too, is a human? From the perspective of the engineer developing an AI (and in Turing's own perspective), a program meeting that standard has won an "imitation game."[1]

Until 2014, no program had ever won.

What happened in the annual Turing trials of 2014? At one level, nothing new: just another chatbot. (That's a program designed to simulate human conversation.) But at another level, and what led to so many breathless headlines: *this* chatbot fooled one-third of its judges.

The much-heralded chatbot presented itself as Eugene Goostman, a thirteen-year-old Ukrainian for whom English was not his native language. This artfully chosen persona served to justify "Eugene's" misunderstandings, awkward phrasings, and follow-up questions. The program's "success," rather than a proof of artificial intelligence, demonstrates natural cunning in human programmers and natural gullibility in one of the trial's three judges.

To be sure, the Turing test isn't, despite its considerable popular cachet, the final word in how to confirm (if/when this happens) the arrival of artificial intelligence. A recent, improved criterion would be the ability to correctly interpret Winograd schemas (first suggested by computer scientist Terry

---

[1] *The Imitation Game*, a 2014 movie about Turing, isn't (despite its title) about AI. It's about cracking the Enigma code.

Winograd). This more nuanced assessment exploits purposefully ambiguous natural-language statements that are readily resolvable with a modicum of (a human's version of) common sense. Here is a simple Winograd schema: "The trophy doesn't fit in the brown suitcase because it is too big. What is too big?"

Perhaps Winograd schemas can't be gamed as readily as the Turing test, but both assessments share a limitation. By either method, any entity that isn't conversant in human culture, affairs, language, environments, artifacts, and modes of interaction with the physical world begins at a serious disadvantage. As a character in my AI novel, *Fools' Experiments* (2008), put it:

> "What kind of criterion was that? Human languages were morasses of homonyms and synonyms, dialects and slang, moods and cases and irregular verbs. Human language shifted over time, often for no better reason than that people could not be bothered to enunciate. 'I could care less' and 'I couldn't care less' somehow meant the same thing. If researchers weren't so anthropomorphic in their thinking, maybe the world would have AI. Any reasoning creature would take one look at natural language and question *human* intelligence."

None of which negates tremendous progress that has been made toward smarter software. In the decades since Alan Turing formulated his imitation game, much has been accomplished. We now have expert systems that, despite their lack of "real world" common sense, apply rules developed by human specialists to focused tasks like stock trading. And a chess-playing program that defeated a world grandmaster. And natural-language processing, like Apple's Siri, and translation software, like Babelfish and its successors, that are no longer (consistently) laughable. And genetic algorithms that "evolve" from crude approximations to reasonable, if not necessarily optimal, solutions to problems. And robots that navigate complex environments, unaided, such as Google's self-driving cars. And "machine learning" software that fine-tunes its behavior in response to the successes and failures of its previous actions. And pattern-matching software that can sometimes identify objects, and hypothesize associations among objects, found within images. And fuzzy logic. And neural nets that simulate human neural tissue. And—

With each such achievement, the realization comes, "Oh, that wasn't a problem of *intelligence*. We just needed to find the right *algorithm*."

Because intelligence, as we humans commonly understand the phenomenon, isn't about algorithms. Intelligence involves awareness and purposefulness. However appropriately a mechanism might respond to stimuli, we resist considering that something "intelligent" when, left to itself, it merely spins its mental wheels.

How and why do you and I initiate actions? What is free will? What is self-awareness? No one knows. Perhaps volition is an emergent property of a very large ensemble of quantum states. Within physics as it is presently understood, only quantum mechanics offers any basis for non-determinism.

Of course, despite almost a century of effort, physicists still fail to agree what quantum mechanics *means*. The plurality opinion of a recent survey of experts was: don't ask. Trust that the math works.[2]

If intelligence does involve volition and volition is indeed somehow rooted in quantum indeterminism, I don't find it surprising that intelligence—whether resident in meat or silicon—remains ill-defined.

Suppose that, someday, humanity's technological toolbox grows to include large-scale quantum computing. Then, perhaps, an entity will arise with a credible possibility of exhibiting intelligence rather than trickery and human mimicry. Won't *that* make for an exciting headline?

If nonhuman intelligences—whether alien, artificial, or both—ever exist, let's hope they have better criteria for recognizing sapient companions than the Turing test.

---

[2] "Why quantum mechanics is an 'embarrassment' to science," Brad Plumer, *The Washington Post*, February 7, 2013, http://www.washingtonpost.com/news/wonkblog/wp/2013/02/07/quantum-mechanics-is-an-embarrassment/.

Neural Alchemist

Tedd Roberts

Movies and TV showed the Zombie Apocalypse as a single event, occurring suddenly due to an uncontrolled infection or some mysterious, mystical event. That couldn't have been farther from the truth. It started slowly, subtly, and we just chalked it up to our own burgeoning medical and technological advancements: a few less patients died, accident victims recovered, once deadly diseases became less so. The First World was caught up in hubris and we patted ourselves on the back for being successful at cheating Death. No one paid much attention to the fact that it was happening in the rest of the world, too. The howling mobs and ravaging hordes would come later... much later.

***

The office walls were a cool, professional blue designed to send the message that this was an office of authority. The University logo dominated the wall behind the receptionist's desk. The occupant of that desk did her best to ignore the man sitting in one of the visitor chairs. Her aura of professional detachment was marred by the furtive glances whenever she thought he wasn't looking.

Somewhere a battery-operated clock ticked loudly in the silence. From an adjacent office could be heard the clicking of keys on a computer.

Professor John Wissen sat waiting.

He has neither comfortable nor uncomfortable. None of that mattered anymore. Nevertheless, he sat.

Waiting.

As if he had all the time in the world.

Tick.

© The Author 2017
M. Brotherton (ed.), *Science Fiction by Scientists*, Science and Fiction,
DOI 10.1007/978-3-319-41102-6_5

The telephone ring was jarring in the near silence of the waiting area. Wissen did not react. Alarm or boredom; neither mattered. The receptionist, however, practically leapt out of her ergonomic chair to answer the phone. After a brief "Hello" and a moment of listening, she hung up the phone, turned to the visitor and said: "The Associate Dean will see you now, Professor Wissen. Please, go right in."

She gestured vaguely in the direction of an interior door, and turned back to her computer screen with a visible sigh of relief when he complied.

***

This office was a distinct contrast to the waiting area, warm beige walls, richly toned wooden furniture, pictures of family members, personal mementoes. The surroundings perfectly suited Associate Dean Laura Diaz. Well-regarded by faculty and administration alike, she had pursued Wissen's case with the administration.

She stood and came out from behind her desk to greet Wissen, shaking his hand – one of the few to still do so.

"Sit, John." She gestured to one of a pair of comfortable chairs, taking the other herself.

"Thank, you, Dean." Wissen said, formally, as he sat. "I appreciate the awkward position this has put you in."

"Nonsense, John." She smiled at him, a heartwarming, genuine smile, not like the furtive glances he'd been receiving lately. "Why so formal? You've called me Laura since we were graduate students."

"Sorry, I just figured with recent events you might need to keep some detachment."

"No, the Board of Trustees specifically asked me to work on this *because* they know we are such old friends."

"Oh. Thanks. I do appreciate it, really." Wissen tried to smile. It wasn't easy. First he had to identify the facial nerve and send it a signal to contract first the cheek muscles, then mandibular muscles and then the skin around the eyes.

Diaz laughed. "John, if you only knew how silly that looks! You are the only person I know that smiles one muscle at a time."

"Technically I'm not a person any more, Laura." Wissen's face fell back into its habitual, neutral expression.

"Well, about that," Diaz continued, "The Faculty Executive Council decided on 'Professor Emeritus' since they didn't think that the Board would go for 'Professor Posthumous'. In fact, the Board agreed, but then they sent the whole thing over to Legal. Once the lawyers work it out, the Board will give final approval." She paused and took a deep breath. "Sooooo… I just got off the phone with Legal. They're having to get pretty inventive, given that there's really no precedent for your situation."

"But they will allow me to continue working?"

"Oh, yes, that was established first, it's the reason for the official position. A liability issue, I was told, if you don't have an official appointment, you can't be here. Your salary, on the other hand…"

"I suppose that means they can still only pay my estate?"

"No, Legal says we can put it in something like a Living Trust, where life partners put all of their assets into a secured fund, but can spend it at need."

"So, I get an Unliving Trust?" Wissen asked, just the sides of his mouth pulled up in another attempt at a smile.

"I suppose we could call it that. Your son will still be the trustee and beneficiary, but you will have unlimited rights at the funds. Legal also says we should pay your apartment and bills from it and register your car to the trust. They've gotten approval to roll your IRA and 403c retirement funds into it as well." Diaz paused and frowned. "The insurance company, though, insists that they won't pay off the life insurance."

"Screw 'em. Tell Legal that if they won't pay Life, then they have to pay Permanent Disability. After all, I *did* die in a covered automobile accident." Wissen's bitter tone belied the blank look on his face.

Diaz laughed. "Actually, Legal told *me* that they could probably get that. If they don't go for a lump-sum payment, it could even cost them more than paying out the life insurance."

Wissen sat in silence for a moment. "But how can Bill administer a trust here? He's in Japan for the next three years."

"Ah, well that's where Legal started getting inventive!" Diaz reached over to her desk and picked up a folded letter. "Here is Bill's designation of a 'memorial gift' to the University. Thanks to his own job and savings, he doesn't feel he needs the inheritance, at least not now. He has authorized us to draw half of the trust as an endowment under the institution's control. He included anything of which he was a beneficiary, such as the royalties on your patents and your retirement funds." She paused. "By the way, what did you do, pour all of Ruth's estate into your retirement accounts?"

"No, actually, most of it went to Bill. It's just that I always contributed the maximum legal amount. It adds up."

"Oh, so that's what it was. I wish my retirement had done as well. Well, with this and the fact that the University still holds your NIH and DoD grants, the Board *had* to reconfirm your faculty appointment and give you back your lab."

"I will *so* enjoy getting out of the basement." Wissen tried again to smile, but he wasn't up to sarcastic yet. The past three months of losing his lab, equipment and students, not to mention car and apartment while living on a cot in one of the antiquated basement labs had worn thin.

"There's still a catch." Diaz reminded him.

"Yeah, I know. I'm still dead."

"Yes, that's true, but I mean your legal identity. Trust aside, you don't legally exist: you can't own property, be paid or enter a contract. The trust will take care of that, but the real issue is identification. You technically can't drive, even though the DMV won't press the issue until your license is due in three years. But you have no *official* status, no ID, no passport, no Social Security number."

"No Social Security, no taxes, no withholding. That doesn't sound so bad."

"No leaving the country, no air travel, no getting stopped by the police, because if you get asked for ID, it comes back as stolen."

"Oh, not so hot then."

"There's one way out, though. If you are right that this is a result of something in the lab, then it's patentable."

"WHAT? You can NOT patent a person."

"Well, technically you're not a person, and we *can* patent a cell culture, and you are most *definitely* a unique cell culture line. Industry Relations has already filed the provisional patent. They just need your notebooks to finalize it."

"DAMN it, Laura, I am not a cell culture!"

"At least it would give you a legal identity."

"Sure it does: 'Property of the University'. Are they going to send Dexter to put a property tag on me? Make me wear it on my forehead, tattoo it on my rump, or just notch my ears like a lab rat?"

"John, it's the only way."

"Sure, Laura, I know. This whole situation is hard; I just never imagined that anything could be worse than negotiating a DoD contract, but this certainly looks that way." He stood up, slowly, and attempted the smile again. "But at least it is something. Thank you Laura."

She also stood, and took his hands briefly, before turning to the office door. "I know it is hard, John, but just look at where you are. We'll make this work."

***

The lecture hall normally seated 125 students; this class had no more than 75 in attendance. Classroom dynamics tempered by medical student politics would ordinary fill up the front two rows and distribute the rest with just a handful of students in the back rows. Today the students were about half in front and half in back. There were only two vacant seats in the back row.

Morbid curiosity or morbid fear? Wissen thought to himself. The bodies in the back row... *Okay, that was a morbid thought...*, the STUDENTS in the back row, plus the unusual proliferation of extra recording devices per student probably accounted for the calls the Dean's office had received demanding that 'The Abomination' be removed from the Faculty.

Just outside the lecture hall was a flyer announcing the special lecture. Some wit had defaced it, crossing out Wissen's name, and replacing it with that of the ghost professor from those young wizard books written a few years back. What was the story? Oh, yes, the old wizard professor had died in mid-lecture and kept on lecturing as a ghost without noticing. Not quite the same, but he supposed there were worse names to be stuck with.

The buzz of voices started to die down. John just sat and waited for his introduction. He was just gratified that students had shown up. There had been a protest piece in the newspaper last week—it was written by the local head of an animal rights group. You'd think that a group that thought a rat, pig, dog and boy were equal could accept someone that was 'differently vital.' But no, they seemed to view it as just another version of human encroachment into the 'pristine realm' of Gaia. The op-ed even claimed that he was disrupting the biosphere by not allowing his remains to "nurture the microbes of the Mother Earth."

The course director was a small man that spoke with great big gestures. With a flourish of hands and arms, he finished the introduction and Wissen stepped up to the lectern and cued his 'wake-up' slide. "Stem Cells. Can't live with 'em, can't be Undead without 'em."

"Stem cells were much maligned in the early part of the century. There was much public outcry over the misconception that stem cells could only come from fetal tissue. People who opposed abortion were afraid that research in stem cells would fuel a need for more tissue, thus encouraging more abortions. Others were afraid of a rash of new cancers or birth defects." That usually got a few nods. It really hadn't been so many years since the government had lifted the total ban on stem cell research.

Wissen had been practicing the lecture with a voice recorder for the last week. Since the accident he'd tended to talk in a quiet monotone. It was an effort to add tone and inflection, but he thought he'd done a decent job. He quickly flipped through several slides showing how most cells in the human body can only form tissues composed of those same types of cells. This was the dull part — the basic background that needed to be covered before getting to the point of the lecture. "We've known for a long time that bone marrow makes a wide variety of very specialized cells; that whole human bodies form from just a few cell types in the fetus. These stem cells have the potential to replace damaged cells in parts of the body that just don't replace all that easily."

He looked around the lecture room. By this point in his career, not to mention the week of practice, lecturing was pretty well automatic. The mouth moved, words came out, but he really didn't have to pay much attention to what he was saying. Instead he looked again at the distribution of students in the room. As expected, the students in the front rows were attentive. The ones in the back looked bored, but a few heads were up and listening. As he went on to describe the many sources of stem cells used in current research: bone marrow, amniotic fluid, umbilical cord blood, transformed endothelial cells, and only very rarely, fetal tissue, he noticed that many of the front row students were writing notes, but the back row students seemed to be distracted or were starting to talk among themselves.

"But what use are we to make of stem cells? Our best example is the brain. For years, scientists felt that a human brain was born with all of the neurons it would ever have — that no neurons could be added or regrown. Now we know that certain areas of the brain, such as the dentate gyrus of hippocampus, have the ability to make new brain cells. Most brain areas do not. What if we could replace the neurons damaged by stroke, injury or disease? Like the old time alchemists trying to turn lead into gold, the Neural Alchemist turns stems cells into any brain cells we need."

There was a stir in the back. Usually by this point in any lecture there would be questions. Medical students liked to gain recognition among their peers by asking questions that they hoped a lecturer couldn't answer. The bragging rights of an unanswered question were a major contributor to student hierarchy. The two people who stood up in the back of the lecture hall didn't look like students about to ask a question, though. For one, the standing male and female did not really look like students; most med students start off with a passing familiarity with personal hygiene and got better once they started performing patient exams. These two looked downright scruffy, unbathed, and wearing dirty clothes. What was that bag at their feet?

"NO ZOMBIES!" The female shouted. The male reached into the bag and threw an elongated object.

It was an arm. A severed human arm. Every face in the hall turned toward the couple.

"UNDEAD. EVIL. NO BRAIN-EATING ZOMBIES!"

The barrage of limbs continued. The man throwing them would certainly not make the big leagues. Most of the limbs were falling in the vacant middle range of seats. A foot made it far enough to hit one seated student in the head. His expression quickly cycled through horror, to revulsion, then pain.

Ah, thought Wissen. *Mannequin parts dressed and painted to look cadaverous.* A few of the students were getting up and approaching the couple, who now turned to exit the room. The movement rapidly turned into a chase, quickly emptying the back rows and part of the front.

"I guess that means today's lecture is done," Wissen told the few remaining students. "Read the assigned chapters and we'll reschedule for next week."

The disruption did have one positive outcome, the next morning there was a petition posted on the student bulletin boards all over campus:

"Got BRAINZ???" It said. "Support Professor Wissen. Support Science. Fight Ignorance." There was an accompanying petition. The Dean's office eventually reported over 1200 signatures. Considering that the Medical and Graduate Schools had 500 students — around 1500 people including faculty and staff, it was a strong show of support. Wissen, however, spent most of the intervening week in a depressed mood, retreating once again to his old basement lab. Only when the rescheduled lecture had full attendance, no disruptions, and a much higher percentage of eager, interested faces, did his black mood start to lift.

***

The lab was cold and dimly lit. Strange that it should be cold in the summer and warm in the winter, but the central heating and cooling conduits had access doors for maintenance on this floor, and they didn't always seal well. The lab had no windows, stained ceiling tiles, broken flooring, leaky water pipes, and uneven pressure in the air, gas and vacuum lines. It was the least popular lab in the building, and was frequently called The Dungeon by the graduate research students. In the months between his 'death' and even after his reinstatement in the faculty, it had been 'Emeritus Professor' Wissen's home and workplace in the basement of the old Pathology research building. Some day it would be renovated, but since an endowment had allowed the University to build a new research laboratory building last year, the renovation had become a low priority.

John Wissen liked The Dungeon. Restoration of his position and funding had been accompanied by assignment of decent research space in the upper floors of the same building, but the disruption at the lecture had convinced him to keep at least *some* research down here and not all in the new lab. Besides, students didn't like the constant breezes from the HVAC and the smell of the vivarium on the same floor. The privacy had allowed him to perform a few procedures out of view of the students.

Research is based on repeating experiments, but how to repeat the singular experience of Undead Professor Johannes A. Wissen, Ph.D.? There was only one source of reanimated tissue, Wissen himself. Obtaining a tissue sample for testing meant taking a piece of his own flesh; and while John was not averse to taking small samples, the process left wounds that did not heal. Healing would have required him to be alive.

He retreated to the office in the back of the Dungeon. At one time it had been used for light-sensitive experiments; thus, once the door closed, there was no possibility of being seen from the outer lab. A large mirror was mounted on the back of the door. John lifted his shirt and stared for a moment at the reflection. His reanimation after the automobile accident had been delayed long enough that the coroner had performed an autopsy. His verdict: Cause of death was cardiac arrest due to rapid impact with a steering wheel. Large incisions started near each shoulder, joined at the center of the chest, then extended down to the upper abdomen, forming a 'Y'. It was stitched closed with precise black sutures, but the edges of the wound remained raw and reddened. Several smaller incisions were not as neatly stitched, marking the sites of previous samples that Wissen had performed himself.

The skin should be cold, gray and necrotic. If I were truly a zombie, I'd look dead, he thought. *Not warm and pink. Not red around the stitches.* There was no sign of bleeding at the incisions, the heart didn't beat, the blood didn't flow, but aside from unhealed scars, he looked as alive as he had ever been.

Today's sample was from the liver. He could get at that though the existing autopsy incision. He unwrapped the sterile covering of his surgical kit. *I don't know why I bother autoclaving it. It's not like I'm going to get an infection.* Using fine scissors, he snipped two sutures from the 'Y'-shaped incision and inserted the biopsy probe. A quick twist captured the liver sample and he removed the probe and placed the tissue sample in a sterile culture dish.

A small drop of dark red blood lingered at the probe site. As he brushed it away to begin resuturing the skin he realized how complacent he had become about the whole procedure. *Damn, I just stuck a whopping big needle into my abdomen without a second thought. I suppose that the lack of feeling — pain or emotion — makes it easier.* Replacing the sutures took only a minute. Looking in the mirror John tightened the silk thread and snipped off the excess with the scissors. As he moved to place the needle in a container for resterilization it slipped out of the grasp of the metal forceps. Reflex born of years of protecting delicate lab instruments caused him to grab at the falling needle. While he succeeded in arresting the fall, the sharp point of the needle jabbed through his protective gloves and deep into the palm of his hand.

Ouch. Wissen thought. *It's just as well that I can't feel that.* John had realized quite early in his new existence that he had very little sensation of touch or pain in his body. That fact had led to the next routine that Wissen performed while he was still alone and in front of a mirror. He pulled over a magnifying mirror similar to the kind used for applying makeup. Using the magnifier and door mirror, he examined each incision, then each patch of unbroken skin for new wounds and injuries. If he didn't want to become a horror movie cliché, he needed to bandage and repair each injury before he risked losing body parts. The new puncture wound in his palm didn't require closing, but it wouldn't hurt to put some tape over it for a few days.

Inspection complete, John exited the office and returned to the outer lab. He would prepare a small sample for microscopy then send the remainder upstairs for the students to culture. Looking at the tissue sample in the sterile dish, he again noticed the drops of blood.

Dark red blood, He thought. *… and that's the problem. Live blood should have gotten redder when exposed to air; dead blood should be dark brown or black.* There had to be oxygen though; somehow oxygen was getting to his brain, muscles and skin without being carried by blood and circulated by the heart.

John dabbed a smear of blood on a slide and looked at it under the highest magnification he could manage on the old light microscope. More sophisticated tools were available in the upstairs lab, but this would do for now. *Red cells, white cells. That's normal. Those filaments, though… It COULD be fibrinogen, except for the fact that they usually show up in clotted blood and as the basis for scabs and scars. None of THAT is happening, so why are they there?* He moved the slide to a new location and adjusted the focus. *Those small cells look very similar to the stem cells he'd been working with prior to the accident. Still, they didn't look quite right, more like immature blood cells.* He'd have a technician run some cellular labeling assays to check it out.

John was not even aware that he had been scratching lightly where he had taken today's sample. Nor did he notice when he started pressing his palm against the edge of the lab bench to relieve the dull ache of the puncture wound.

***

One week later John again entered the privacy of the downstairs lab. The recent liver tissue and blood had indeed included stem cells, along with more of the filaments — not just in the blood, but in the liver sample as well. In order to start sustainable cultures, he'd need a larger tissue sample and considerably more blood.

He was facing away from the mirror as he removed his shirt. As he turned around and reached for the sampling probe he stopped…

…and stared.

The incision immediately over the prior sampling site was closed. About an inch of new scar tissue had formed in the middle of the autopsy incision.

I guess I'll have to go in from a different site. He began to snip away sutures below the new scar and prepared to insert the slightly larger sampling probe. *DAMN. That HURTS!* He retrieved the probe and sample, but had to sit down and rest before attempting to suture the incision. He might have to try some anesthetic before collecting the blood sample.

Out in the lab he found an anesthetic spray used to desensitize incision sites during animal surgeries. The suture sites burned from the slight punctures of the needle, but the spray relieved enough of the sensation that John could consider the next step.

With no heart beat or blood circulation, it would not be possible to just stick a needle in a vein and draw blood. He had planned to make a longitudinal incision in a large vein near the ankle, and rely on gravity and pressure on the calf to squeeze enough blood into a test tube for culture. A quick test of the scalpel on the skin of the ankle revealed no sensation down there — yet. Still, he was reluctant to cut on himself and repeat the experience of the biopsy probe. This would require some assistance.

Phil Wohlrab had been a friend and occasional co-worker since college. John had spent a few years working in various labs before going to graduate school; Phil had joined the Army, become a medic, and then went to medical school after being discharged. John had just joined the faculty when Phil arrived as a first year Internal Medicine resident. They'd rekindled their friendship and become like brothers, even to the point of helping each other through the pain of losing spouses. Most recently, Wohlrab was working with the Aging Center to address problems of administering medications to the elderly populations. He had extensive experience with patients having collapsed veins, so John called him in to assist.

The basement lab seemed crowded with John, Phil and Laura Diaz in it. John was seated in a reclining chair while Phil inserted a cannula into one of the large veins in John's neck.

"This is no different than a central line, John. I'll insert the tubing far enough that it should be at the right ventricle, then draw blood."

"Urgh," was all John could manage. Phil had placed a high collar around his neck to keep it in the appropriate position for the procedure.

"I think that was 'Thank You'," injected Laura helpfully.

"No," gasped John, "that was 'Hurry up'."

Phil drew 10 cc's of blood and then quickly removed the tubing. The blood in the syringe was dark reddish-brown, but the drop that formed at the entry site on the neck was a brighter red.

"Definitely oxygenated blood, John," said Wohlrab. He transferred the blood into a tube containing chemicals to preserve the sample, and then placed the tube in a bucket of ice chips. "Do you want me to take any other samples while I'm at it?"

"No," said John, removing the collar and massaging the neck muscles. "Either this is it, or I'll have to submit to a full surgical procedure to get all of the samples we'd need. Since I have no desire to repeat that autopsy, this had better be it. Thanks, Phil, you're a good friend."

Laura spoke up. "Did you know you actually *smiled* when you said that, John?"

***

The International Union of Pathology and Pathophysiology was being held in Innsbruck, Austria. The Congress Centre was a bare half kilometer from the old city and former residence of Holy Roman Emperor Maximillian. The juxtaposition of old and new was never more apparent than in the modern teleconferencing facilities of the Salon where Professor Wissen was scheduled to address the conference. The legal issues had long since been resolved to the point where he could have travelled to the scientific meeting, but an excess of publicity coupled with recent developments made it safer to address his colleagues over a closed circuit television link.

The university's teleconferencing studio was a strange mixture of television news studio and academic office. John sat at a desk in front of a video camera; in front of him, two video monitors showed the assembled scientists and the master of ceremonies beginning the introduction in Innsbruck. To one side of the desk was a computer which would control and display the presentation simultaneously in the local and remote locations. On the wall behind him were the University Medical Center logo, a white board, and a bookshelf with books arranged to prominently display key Pathology textbooks.

In Austria, the speaker was finishing the introduction: "...and without further delay, I present this year's Keynote Lecturer, Professor John Wissen." On that cue, John tapped the computer keyboard and started playing a video that had been prepared over the previous months in anticipation of this presentation. It started with him seated in this very studio, addressing the camera:

"Fellow Scientists. I won't dwell on the sensationalism and lurid background of this finding, but I am here to report that our research team has made an astounding discovery in stem cell research. For years we have known that life is an intricate balance of metabolism and diffusion. The mammalian physiology consists of a closed circulatory system that supplies individual cells with oxygen and glucose for their individual metabolic needs, and removes the organic wastes provided by those same metabolic processes. But what if we could remove the redundancy of identical chemical processes in each cell and simply supply the energy through a distributed network between cells? Individual cells would not metabolize, nor would they excrete, but they would all receive exactly the energy they needed in order to function.

"We have determined that specific differentiation of stem cell line UMC325 into a novel cell type that we call UMC325.JW provides just that function. JW cells consume oxygen and organic molecules and transfer the essential energy storing molecule — ATP — directly to any mammalian cells. JW cells are highly mobile and quickly permeate living tissues, leaving an interconnected matrix in their path. This matrix makes blood circulation, and even a beating heart, unnecessary. My unique existence is because of JW cells."

John watched the audience on the monitors as the video continue. There was much nodding of heads, whispered comments, and furious note-taking. The video continued with time-lapse recordings of the essential experiments that proved the thesis. Laboratory rats were injected with the JW cells, 24 hours later their hearts were stopped by electric shock. EEG and EKG monitors showed no activity for 30 minutes. To all appearances, the lab rats were dead. Between 30 and 60 minutes after the heart shock, each laboratory rat began to twitch, move its limbs, and eventually get up and walk around. EEG tracings revealed renewed brain activity even though EKG showed a complete absence of heart beat.

The video proceeded to show repeated demonstrations with cats, dogs, and monkeys. *And that was when the real problems started*, John thought. Organizations that fought to prevent animal "death" in medical research were strangely unsympathetic to the fact that those same animals were brought back to "life" in Wissen's lab. The protests and death threats had caused him to move out of his apartment and take up residence again in his basement lab. At least the university had furnished it for him this time.

It's a good thing no one ever saw the final step in the progression.

Once it was determined that JW cells already present could return an animal from death, John knew that he would need to test whether they would be effective if administered to a creature — a human — that was already dead. The problem was that legally, ethically, he could do nothing of the sort. The Buckley incident was a rare accident, but it had nearly cost him everything.

Joseph Buckley was a computer tech for the university. Three months ago, he'd had the unenviable task of babysitting the uninterruptible power supplies serving the computer room while University Engineering repaired the emergency power switch for that building. Despite massive battery backups, each server was connected to the emergency power circuits to ensure that no data could be lost due to power interruptions. During the switch repairs, Buckley had to make sure that the servers were running off of the batteries, or carefully shut them down with their data intact. The power surge was unintentional, but it caused one UPS to fail catastrophically. Fortunately for the computers in the room, the resultant arc found a closer path to ground; unfortunately for Buckley, his body completed that circuit.

John was enroute from the basement to his upstairs laboratory when the lights flickered and died. He heard the scream from the computer room and was the first to arrive at the scene. When he discovered that Buckley had no pulse, he realized he had to do something. While he could not administer 'mouth-to-mouth' resuscitation — after all, he had no 'breath' to share — he could at least provide heart compression until the paramedics arrived. After nearly an hour with no success, the paramedics declared Buckley dead and sent his body to the morgue. When Joe woke up screaming in a morgue drawer six hours later, all Hell figuratively and literally broke loose.

Without injection, without ingestion, and without any overt intent to use Buckley to test John's theories, somehow the JW cells had been transferred. The Ethical, Legal and Scientific Investigation Board determined that the transfer had occurred during the prolonged skin-to-skin contact during the CPR attempt. John had opened Buckley's shirt and placed his hands directly on his chest, providing a pathway for JW cells to migrate from Wissen to Buckley. Since that event, John had been in effective isolation and anyone in contact with him during that past months was subject to intense examination and distrust.

The video presentation was drawing to a close. On the computer screen, John's recorded image was talking about the unique metabolic requirements of JW cells. A red light flashed 5 times in the studio, and the "live video" indicator was displayed on all video screens.

"That video was prepared over the previous three months. You can now see why I was unable to attend the meeting in person." The audience members were visibly startled by the biological isolation suit, not to mention John's gaunt and withered appearance compared to the recording they had just viewed.

Wissen coughed, and resumed in a hoarse voice. "We have since learned that JW cells infiltrate brain, nervous system and muscle within hours. However, they do not appear in endothelium or the lining of the gut for many months. Until the digestive system is reactivated, the subject is unable to process or absorb nutrients from food. JW cells have a very efficient metabolism, but eventually the need for nutrients causes them to break down the very cells they have reanimated.

"For those who are encouraged by these breakthroughs, I must caution you that they are short-lived. For the many who are outraged and offended at my very existence, I can likewise assure you that it will soon end.

"I caution you, though. I may have been first, but I suspect that I will be far from the last of my kind. Knowledge, once found, cannot be undiscovered."

The video camera turned off, and the computer screen displayed a message that Professor Wissen would be unable to answer questions. Further inquiries should be directed to office of sponsored research at his university.

***

Laura Diaz had been sitting out of camera view. She quickly rose and came to Wissen's side to move his wheelchair as soon as the video feed was turned off. "John. You're weak. Please eat; we both know what you need."

Wissen looked at his oldest, dearest, and possibly last remaining friend with sadness. "No Laura. There are some things that even a renegade scientist can't do, zombie or not."

"Yes, you can John. This is a medical center. There are ways. It does *not* have to be like this. It's the *one* thing we learned from Buckley."

"Oh, yes. That will go over well. The Vegan Revolution won't even let the native Scots eat haggis. No, I've had my chance. I can't do this anymore." He refused to look at the tears in Diaz' eyes, but he took her hand and she said nothing.

The JW cells found in Buckley were much more developed than the unpurified cells which had originally accidently infected Wissen. Buckley woke up in six hours compared to John's seventy-two. Joe's second life lasted just two months to John's twenty-eight.

Of course Buckley started having food cravings only 4 weeks after his reanimation. The scientists would never know just how long he might have existed if he'd had continued access to food.

I'll be damned if I let that happen to me. Buckley had been literally dismembered by an angry mob after he'd been discovered bent over the bloody body, eating the heart of his latest victim. *Then again, I'm probably damned anyway. After all, I've lived more than two years in Purgatory, if not outright Hell.*

"Laura, promise me that you'll figure it out. Either figure it out, or destroy the JW cell line once and for all."

"Yes, John, I promise."

"Good. I would really rather not be remembered for unleashing zombies on the world."

Laura looked shocked. "Surely you don't think…"

Wissen sighed, and it sounded like a death rattle. "I do, Laura. 'Not with a bang,' nor a whimper, but a starving bloody madness. I just hope we're not too late."

***

But they were.

Afterword

Neural Alchemist is the result of several challenges imposed upon me by friends who are SF authors and readers. Frankly, the effort started with the joke "Professor Posthumous," based upon professors of my acquaintance who keep working in their chosen fields until forced to stop… but never by choice. Soon after, the first challenge was imposed by a group of fledgling writers of which I was a part. We were instructed to write an opening scene using 'show, don't tell.' The image of Wissen outwaiting the office clock was the result, along with the Dean's meeting and the inclusion of my private joke.

The second challenge was of writing outside of my comfort zone. As a scientist, if I follow the instruction to "write what you know," then obviously, I would write 'exploration' stories about scientists, in labs, or in academic settings. I once told a very good friend (who writes SF, fantasy, space opera, mystery and historical fiction) that I did not feel I could write in a fantasy setting. She replied "when asked to write outside your comfort zone, do not say 'No,' say 'How many words, and when is it due?'" She then challenged me to write a three fantasy short stories: one each about a zombie, a werewolf, and a vampire.

The third challenge was from the writing group: Take the opening scene, follow my friend's advice, and finish this unusual scientific fantasy. I confess that I could not quite leave out the research scientist, but I will justify the finished product by observing that the line between science and fantasy is getting ever more blurred with recent advances in stem cells, gene editing, quantum physics and virtual reality.

The setting of this story is very real. The offices, labs, classrooms and auditoriums are all modeled after places I have worked or visited. The tissue engineering and stem cell research group is in the building next door to my own. Not that it's my field, mind you, but I have worked with researchers on testing stem cell transformation into precursors of the neurons — brain cells — that are the subject of my own research. I have taught nerve cell physiology to their students, and in fact, one of their professors was a classmate of mine in graduate school way too many years ago. On that note, I have not transplanted any of my colleagues into the story. The fictional characters and names are... mostly... made up. One of my close friends is included with his permission, and some readers may recognize the inside joke among some authors of 'killing Joe Buckley.'

Do I necessarily think that stem cells would result in a zombie plague? Not really, stem cells require very careful growth conditions; notably, they require specific trophic factors to transform them into the various specialized cells comprising the human body. The concept of stem cells taking over the oxygen and energy transport function by diffusion without blood circulation is clearly a McGuffin or gimmick requiring the 'willing suspension of disbelief.' Even if stem cells *could* utilize the scaffolding and chemical residue of a cadaver to grow, differentiate and repopulate the body, too many specific connections and functions would be lost. In the brain, our memory, skills and personality are a product of the synapses and connections between neurons. As we grow and learn, we change those connections in order to store new information. These connections are fragile, however, and even the process of remembering can produce subtle changes in the stored pattern. Unfortunately, the neural connections are among the first to be lost in death or serious head injury. The likelihood of restoring them via stem cell regrowth is very low, but may perhaps be responsible for a scientific version of the 'brainless zombie' of movie and TV cliché.

Finally, I have attempted to capture the cycle of life-to-death and recycle it from reanimation-to-decay because my own research in tissue physiology suggests that no 'cell culture' version of reanimation would last forever. Even if stem cells completely replaced the deceased cells of a cadaver, once they run out of cells to replace, they run out of raw materials and nutrients for sustenance. Only an additional, easily metabolized source of nutrients would stave off eventual decay. Again, the fiction cliché of the peculiar dietary needs of zombies actually makes a little more sense from this scientific standpoint. A reanimated corpse would not have a functioning digestive system and liver; thus, liver and brains, with their complex proteins, sugars and lipids, would be the perfect nutrient, requiring less metabolic processing than muscle or vegetable matter.

This is a fun exercise, and if a few readers go out and learn a bit more about stem cells and neural function, my task has been accomplished. I firmly believe that good 'hard SF' is a means of scientific outreach that is largely overlooked as we scientists seek to educate and inspire our successors. I am grateful to have had the chance to contribute.

Hidden Variables

Jed Brody

Anything worth saying, is worth saying again. So here goes.

I found the hidden variables, the ones that Einstein always insisted on. It turns out they were under my couch all along, covered with dust and the black and brown hairs of a small yappy dog that died three years ago.

I haven't told anyone about the hidden variables yet. Nothing's really changed for me or for anyone else. I rise at dawn and swim in the ocean, I go to campus to teach my class for an hour, and then, through the magic of tenure, I depart immediately and hike in the forested hills. On Tuesdays, Thursdays, and Saturdays, I attend karate classes followed by advanced sparring with the other black belts.

I want my twin sister Chloe to be the first to know about the hidden variables. She was the first and only person I ever told about my mystery poems. I was only six when I began finding the poems under rocks, folded in the pages of books, and sometimes even in my pockets. I didn't know this was unusual; I thought this happened to everyone. Like the time I went to my friend Ali's five-year-old birthday party and her mom wasn't there, and I said, "She must be at the observatory or at yoga." Because, obviously, those are the only possible reasons for a mom not to be at home.

It was Chloe, who was also only six, who had the insight that mystery poems were not common occurrences, and I should keep them hidden or

© The Author 2017
M. Brotherton (ed.), *Science Fiction by Scientists*, Science and Fiction,
DOI 10.1007/978-3-319-41102-6_6

claim authorship. It's cute to have an imaginary friend who leaves you poems, but not when some of the poems are dark and apocalyptic. If such poems are found around you, it's better to say you wrote them on a whim, as an adventure, out of curiosity, just an exploration; than to say they were left for you by some kind of unidentified invisible paranormal entities.

And, of course, the most obvious explanation is that I *do* write the mystery poems but suffer selective amnesia about it. The main reason I don't like this theory is that if my subconscious mind is writing poems, I wish it would do a better job. They're really not that good.

Chloe and I speculated long and hard about where the mystery poems were coming from. Was *she* writing them but denying it? Did I write them in the future and travel back in time to deliver them to my younger self? Were they written by a secret (and totally creepy and inappropriate) admirer? Ultimately, we had to give up and accept that the mystery poems were just an unsolvable riddle.

Chloe is coming over today to celebrate our 33rd birthday. She hasn't visited for almost 15 months. She's been traveling through Asia on a quest for transcendental experiences. I yearn to see her again, but I'm also apprehensive. It's the apprehension that precedes looking in the mirror. You never really know what you're going to see, and you're not sure if you're going to like it.

I lift a small teapot from the two-tiered, glass display-table where I keep all of Chloe's pottery. The teapot has always been my favorite piece, painted with psychedelic teal and pink swirls. I'm not sure what kind of tea it was meant for, but I prefer Earl Grey. I set the clay teapot on the coffee table and start water boiling in a mundane teapot on the stove.

I return to the display table and examine the teacups. None of them match, so I pick a tall, blue, ridged cup, and a squat green cup on which lambs are painted. When I lift the lamb cup, I discover a small piece of folded paper underneath. A mystery poem! I consider waiting for Chloe before reading it, but then I quickly unfold the paper.

The sky's the blue of Krishna's breast
The sun devours mass a billion of me per second
Once I heard of a black hole that devours thoughts
Oh windflung plank, oh grazing bee, into that tempest lead me

I smooth out the paper and place it in my folder of mystery poems. I set the folder on the coffee table, next to the teacups and teapot. I open the folder again and look at the one item that's not a poem, but is the greatest mystery of all: the six-month ultrasound image, where Chloe appears for the first time. In all the earlier ultrasounds, I was alone in the womb.

I hear the mundane teapot whistle just as Chloe rings the doorbell. She always shows up at the right moment, when everything's ready. No reason to sit through the primeval chaos of the first two trimesters if you don't have to. I turn off the stove and then open the door.

Chloe's black hair is shorter than ever, but her smile is more radiant, to an almost unimaginable degree, and her muscles are even more defined. I squeeze her in my arms and marvel at the resting power in her shoulders and back.

We pull apart and smile. "Happy birthday!" we shout in unison. Then we giggle like when we were eight.

"Come on in!" I say. "You look great!"

"You too," she says, flopping onto the couch. "Tenure's doing you good."

"The most ironic thing about tenure," I say, pouring boiling water into the clay teapot, "is that students are giving me much higher course evaluations, now that it doesn't even matter. I think, when I was under the stress of working toward tenure, I passed the stress onto my students. I like to say that I was stress-passing on them. To remind myself not to do that, I put a No Stresspassing sign up in my office."

"No Stresspassing. Very nice," Chloe says with a grin. "I wish my pottery professor had observed that rule. We could tell when he was fighting with his wife because he'd hurl students' pots against the wall. He said he wasn't there to kiss our boo-boos and pretend that we had talent. He was there to be a merciless destroyer of the untalented, the unoriginal, the uninspiring and uninspired. He made students cry."

"He didn't smash any of your work, I'm sure," I say, sprinkling loose tea leaves into the steaming water.

"Actually, he smashed some of my best pieces," Chloe says, shrugging and briefly rubbing the mole above the left corner of her mouth. I catch myself starting to rub the mole above my mouth, on the right. "But whether I sell them, give them as gifts, or have them destroyed by an emotionally disturbed mentor, I don't get to keep them."

"Have you been doing any pottery lately?" I ask, settling into my easy chair.

"Not lately. It's hard to make a living off it, and if you're not making a living off it, it's an expensive habit. And I have too many expensive habits as it is," she grins.

"Such as scouring the world's largest continent for the secrets of spiritual ecstasy, unwavering bliss, and other altered states?"

"I was thinking of organic almond butter," she laughs. "Have you seen what the drought has done to the prices? But yes, I have sought after wise old men. Wise old women, too. They're wiser but harder to find."

"Have you learned anything?" I ask, pulling my feet onto my chair and adjusting my pink socks.

"Yes," she says, smiling, but leaning forward earnestly. "First, there's no one true path. Second, for me at least, spiritual knowledge is worse than useless. Knowledge without practical techniques is itself a technique, for frustration and depression. I took Buddhist philosophy in college. On the first day of the semester, the professor talked about how her students were transformed by the class. Runners ran faster, fencers fenced better, watermelon-seed spitters spat watermelon seeds farther. So I expected to get better at everything. Everything!

"But by the end of semester, though I'd learned a fair amount, nothing else had changed. And my failure to transform made me *feel* like a failure, a fool, a frump."

"A fart?" I suggest.

"A forsaken fuck-up," Chloe laughs. "But what I've found, finally…"

"…felicitously…" I interject.

"…fortuitously," she resumes, "is that practical techniques to feel the life force, are what's needed to experience the elevated states described in the scriptures. Practical techniques of meditation, yoga, breath. Prayer, if you prefer. The resulting sensation of life force *is* the meaning of embodying heaven on earth, it *is* the meaning of being an empty vessel for a greater power, it *is* the meaning of aligning the ego with the universal self. There is no need, except for recreational purposes, to study the *meaning* of any of these concepts. The only need is to *feel* them as concrete bodily sensations.

"The great error of Western civilization was Descartes' 'I think, therefore I am.' The reality, which is a blessing and a responsibility, is, 'I feel, therefore I want to cultivate feelings of bliss.'"

"That's lovely," I say, lifting the teapot, "and do you know, there's something I've been waiting to tell you—"

"A mystery poem!" she exclaims, pointing under the teapot. Her eyes widen with glee.

"That's my second one today," I say as I pour her tea. "I don't think I've ever gotten two in a day before."

"May I read it? Yes?" she asks. She unfolds the pale green paper.

What You'll Find Among My Ashes
Endless longing and endless fulfillment
The profound peace of the M5 globular cluster (it's a galaxy)
All the sugar donuts I ate when I was seven
All seven oceans

The crash, rumble, and hiss of waves
The spray of salt
A snow-encrusted buffalo
The roar of lions

We sit grinning at each other.

"You're sure you didn't slip that under the teapot with your crazy ninja skills?" I ask.

"I didn't," Chloe says, "though I do have crazy ninja skills. But what have you been waiting to tell me?"

I open the folder of mystery poems and take out the six-month ultrasound.

"Aw, we're so cute," she says.

I smile. "I accidentally dropped this under the couch last week. And when I reached under the couch, I found the hidden variables. The hidden local variables that Einstein insisted on."

"Although I do have supernatural powers," Chloe says, "knowledge of physics isn't one of them. So you'll have to explain that to me."

"I did, when I first learned about quantum entanglement, when I was an undergrad. Are you sure you want to hear it again?" I ask.

"Anything worth saying, is worth saying again."

I laugh. That's what she said when we were little, and she fell asleep as I told her bedtime stories. She'd wake up and ask me to repeat what she'd missed. I was irritated, but she said, in her adorable baby voice, "Anything worth saying, is worth saying again."

"Okay," I say. "There's a particle called a pion."

"As in, I hope I don't get *pie on* my new dress," Chloe says. "Yes, I remember the name."

"Right. So, the pion spontaneously decays into an electron and a positron, which fly away from each other. These particles have a property called spin, which causes them to be deflected in a certain kind of magnetic field. If a particle is deflected upward in a particular magnetic field, let's call it spin up. If it's deflected downward, we'll call it spin down. The original pion had no spin, and this means that if the electron is spin up, then the positron must be spin down, and vice versa. As soon as the electron's spin is measured, the positron, no matter how far away it is, is compelled to have the opposite spin. Einstein hated this idea. He called it 'spooky action at a distance.'"

"Am I supposed to be amazed by this?" Chloe asks. "How's this any different from having a red marble and a green marble in a bag? I reach in without looking and take a marble in each hand. When I open my hands, am I supposed to be amazed that the two marbles are different colors?"

"That's basically what Einstein said," I say. "You're saying that the two marbles had their distinct colors all along. At first, you're simply ignorant of which marble is in which hand, and when you open your hands, the marbles don't change. The only thing that changes is your knowledge of which marble is in which hand.

"A variety of factors determine which marble ends up in which hand: the original locations of the marbles in the bag, the way you shake the bag before you reach in, the way your fingers grope sightlessly through the bag. If we had enough information about these factors, we might even be able to calculate which marble ends up in which hand. This information that we don't have is called 'hidden local variables.'"

"And that's what you found, under your couch?" Chloe says, raising an eyebrow.

"Yes!" I say. "Uncanny, isn't it? But I still haven't explained what's amazing about the spins of the electron and positron. In the marble experiment, when you open your hands, the marbles don't change. If you see the red marble in your right hand, you infer, correctly, that the red marble was in your right hand all along. The observation of the red marble did not transform it into a red marble. You then know, without even looking, that the green marble is in your left hand. But the green marble, too, was there all along. The observation didn't make it green. We might say that hidden variables determined which marble ended up in which hand.

"In quantum physics, however, there's the claim that the electron and positron did not have their spins all along. Before either particle is observed, their spins are undetermined. Each particle is in a kind of undecided state. It's not that we're just ignorant of their spins before we observe them; the spins themselves are inherently undecided and unknowable before we observe either particle. The observation is not passive; the observation transforms the particles.

"So, there are two things that are supposed to amaze you. First, the observation of a particle determines its spin—a spin it didn't previously have. It previously had latent possibilities of spin. The observation selects the particle's spin from two latent possibilities. Second, the observation of just one of the particles instantaneously selects and determines the spin for both particles. The observation of the electron determines the spin of the distant positron."

"Okay," Chloe says, sipping her tea, "I now see why this is amazing, if it's true. But how can you possibly know that your observations are changing the particles? You certainly don't have any evidence about the particles' spins before you observe them. The only way to get any evidence is through observations."

"This is where Bell's theorem comes in," I say. "Someone named Bell said, suppose the particles really have their spins all along. We don't know their spins yet, but their spins are determined through some unknown factors, which we call hidden local variables. Bell showed that any possible choice of hidden local variables must contradict the predictions of quantum physics. But experiments agree with quantum physics, so therefore experiments contradict hidden local variables."

"And yet you found the hidden variables, under your couch?" Chloe asks.

"Yes. But I still believe in quantum physics. It doesn't really make sense, does it?"

"No," Chloe says. She rotates her mug on the coffee table, making a soft scraping sound. "How do you know you found the hidden variables?"

"That day I looked under the couch," I say, "I saw our six-month ultrasound, as well as a lot of dust, dog hairs that I'd never vacuumed up, cracker crumbs, and sesame seeds. Then I saw a churning purple mist which had no reason to be there, but for some reason I wasn't scared. The purple mist spilled out from under the couch and entered the soles of my feet, rising up to the top of my head. And then I knew everything."

"Everything?" Chloe says.

"Well, everything I turned my attention to. I saw red lava bursting from the bottom of the ocean, shooting out jets of steam while blackening and hardening. I saw mammoth tusks, megaliths, and stone altars reddened with blood. I saw galaxies collide and rip each other apart. I can look anywhere in the past, present, or future, but I turn away from the future because it freaks me out. I can see lottery numbers, but I don't need any more money."

"That's really exciting, to say the least," Chloe says, stretching her arms and rotating her ankles. "But I don't think you found the hidden local variables."

"No?" I frown, gnawing my lip. I feel unreasonably irritated and strangely alarmed. "But that's part of this knowledge, knowing that the knowledge, itself, is made up of all the hidden variables, all the factors that determine everything that is."

"Yes," Chloe says. She smiles, but sadly, and she sweeps her fingers across the mole above the left corner of her mouth. "But you didn't find the hidden *local* variables. You found the hidden *global* variables. The hidden variables that don't contradict quantum physics. The hidden variables that permit spooky actions at a distance: a positron's spin depends not only on local conditions but also on the measurement of the distant electron."

"That's right," I say, narrowing my eyes in concentration. "If hidden variables determine a particle's spin based on something far away, there's no contradiction with quantum physics. Bell's theorem is limited to hidden *local* variables."

"So you can have your cake and eat it too," Chloe says with a maddening smirk. "You found the hidden variables, and you preserved the implications of quantum physics."

"But how did you know this?" I ask, feeling an inexplicable harrowing dread. "How did you know about hidden global variables?"

"Because they're mine," Chloe says evenly, unblinking.

"What do you mean?" I gasp, but I can already feel the knowledge slipping away. The vast realms within my awareness shrink, they crack and splinter like glass, they recede like a tide, leaving behind parched sands and sun-bleached shells.

"They're mine now," Chloe sighs.

"You can't have them!" I shout. "I found them first!" I can almost see purple mist pouring out of me and gushing into Chloe, as though she were a vacuum cleaner.

"Give it back!" I yell, and I leap over the coffee table to tackle Chloe. But she somehow anticipated my move, and I slam into the vacated couch. I spin around and see her slowly pacing around the coffee table.

"This will be easier for you if you relax," Chloe says, swinging her arms gracefully as though wading through water.

"Thief!" I shout, clenching my fists and springing to my feet. "Hypocrite! Is that what years of spiritual searching has brought you to? You put on a great show of cultivating inner peace, but you steal from your own sister?"

"Which of us is behaving more peacefully right now?" she asks, languidly bending a knee and circling one foot in the air.

"Don't turn my words against me! How dare you do this to me!"

I lunge and swing my palm towards her face, but she somehow ducks under my arm and trips me. I stumble into the display table. Her pottery jitters. A small vase tips over and rolls onto the floor, but the carpet saves it from shattering.

"A karate strike," she says. "How cute."

"It's not cute!" I say, and I throw a punch, which she deflects. "I can break a stack of two-by-fours with my hand!"

"Like an angry child throwing a temper tantrum," she says, continuing to parry my strikes. "Cute."

I soar toward her with a leaping spinning hook kick, but she dodges and slaps my foot in the same direction it was going. Losing my balance, I land on my back and break my fall by slamming my arms downward. My right arms catches the corner of the coffee table. It flips over, hurling the teacups and teapot against the wall. The teacups shatter, but the teapot falls onto the carpet, dripping and unharmed.

Chloe squats deeply on one leg and extends the other leg fully along the floor. A ridiculous pose, an easy target. But when I descend on her with blows, she evades me again.

"Karate is no match for taijiquan," she says placidly, shifting, receding, always just out of reach. "The Supreme Ultimate Fist. According to one account, invented by a general for people who were already experts at the martial arts."

She again deflects my fist, causing me to punch several of her pots off the display table. Clay shards are beginning to litter the carpet. The faster I strike, the more forcefully Chloe redirects my limbs.

"It's not fair!" I say, starting to pant from exertion, but I don't pause in my attack. "I do all the work, and you have all the fun! The whole time I was working, you were on a dream vacation to temples and mountain hermitages! And now that finally everything is perfect for me, you come and take it all away!"

Chloe gazes at me imperturbably, almost pityingly. Her arms are a blur as she deflects my strikes. I feel her fingertips like hailstones against my arms and legs.

"It was inevitable that we'd fly off in opposite directions," she says. "One to the east, one to the west. Opposite sides of the globe. Our heads pointing in opposite directions. While the sun in your sky sets, mine rises. As smoke ascends, snow falls. Now I am in the season of blossoms, and you face your day of autumn."

"No!" I scream, and I hurtle toward her with my strongest punch, a lethal punch. But, predictably, she whisks herself out of the way at the last minute. I crash into the display table, which topples. The two glass table-shelves shatter, along with all the clay artwork. I steady myself, try to catch my breath, and begin to circle Chloe around the coffee table.

I must have hit my head at some point because my vision dims and returns. I have the sense of circling faster and faster, though I hardly feel that I am moving. I glare at Chloe and gasp.

"Your hair!" I shout. "How did your hair get long again?"

But then I see it's worse than that. The mole is over the right side of her mouth. She has my mole. She has my hair. She's me.

The room is spinning even faster, though Chloe—in my image—is standing still.

"Once we were two, but now we are one," Chloe says.

She lifts the six-month ultrasound from the wasteland of broken ceramic. The room is spinning so fast that all the colors are blending into white.

"Oh look!" she says, pointing to the floor that I can barely see. "A new mystery poem, which was hiding under the ultrasound! That's three in one day!"

But I'm not looking at the mystery poem. I'm looking at the ultrasound. And though I can't see anything else clearly, the ultrasound image is immaculate. And only one fetus is visible.

"From now on, there is no Chloe Elghenyin," Chloe says. "There is only Stephanie Elghenyin. So it is not you who vanish. It is I."

Chloe—me—whoever she is, raises her arms as though summoning spirits. Tears streak her cheeks, and her face is clenched in sorrow.

"I'll miss you," she says. "Though soon I, and everyone else, will remember the last 33 years differently from what we remember now."

The room fades to white, which suddenly turns to black. I am alone in blackness. Then I hear Chloe's voice.

"Here's your final mystery poem," she says.

> *I was a rock, a willow, a moon*
> *I waited for the rain to fall, but I was the rain*
> *I was a snow leopard, a heron, a nightingale*
> *I watched until the thundering waves shattered skyscrapers, but I was the waves*
> *Because the reason spiders eat their young is neither cruelty nor hunger*
> *But yearning for reunion with their offspring*
> *So the rear of my funeral pyre is the red giant sun when it*
> *Becomes the sky*
> *And expands past the orbit of Venus*

I float in the dark, sullen, alone, and only a little bit scared. I hear a faint buzz and see purple mist rising everywhere. It has come to carry me to the realm of unwavering bliss and unending tranquility. But I know this mist. It permeated me for a week, so I know that it permeates all things and all times. So I direct my attention to a time 33 years and three months ago. And I ask, and receive, permission to go.

Hello Stephanie.

Good to see you again.

Thought you were alone in here?

We're all soft little arms and faces in a warm dark salt bath. Cozy and cuddly.

Anything worth saying, is worth saying again.

Afterword

A lot has been written about quantum entanglement. My favorite is *How the Hippies Saved Physics* by David Kaiser.

I'd like to work through an example adapted from *Quantum Non-Locality and Relativity* by Tim Maudlin.

This example is based on experiments with light, which is made of photons. We can shine light on something called a polarizer. Each photon reaching the polarizer will do one of two things: it will pass through, or it will be absorbed. The probability that a particular photon will pass through depends on the orientation of the polarizer; we change the polarizer's orientation simply by rotating it. Let's say that we're interested in orienting a polarizer in one of three ways: horizontally, 30° above the horizontal, or 60° above the horizontal.

Suppose we create a pair of photons that go in different directions (left and right). The photon on the left goes toward a polarizer (oriented either horizontally, at 30°, or at 60°), and the photon on the right goes toward a different polarizer (orientated in one of the same three directions). It's possible to create pairs of photons about which the following observations are made:

Observation 1.
If the two polarizers have the same orientation, the two photons always do the same thing: Either they both pass through the polarizers, or both are absorbed by the polarizers.

Observation 2.
If the angle between the two polarizers is 30°, the two photons do the same thing 75% of the time. 25% of the time, one passes through its polarizer, while the other is blocked.

Observation 3.
If one of the polarizers is horizontal and the other is at 60°, the two photons do the same thing 25% of the time.

This is based on real observations; this isn't made up.

We want to determine whether the photons have their polarization properties all along, or whether the photons are in some kind of mysterious indeterminate state prior to reaching the polarizers.

Let's assume that the photons have "hidden properties" prior to observation; observation merely lets us view the properties that the photons had all along. Let's see where this assumption leads.

The hidden properties of a photon might be this:

(Would pass through a horizontal polarizer. Would be blocked by a 30° polarizer. Would pass through a 60° polarizer.)

It took a lot of words to write that. Let me represent *exactly the same thing* in abbreviated form:

(Horizontal → Pass. 30° → Block. 60° → Pass)

If one photon in a pair has these hidden properties, the other one must have the same properties. To understand this, suppose one photon in a pair has the hidden properties listed above, and the other one has these properties:

(Horizontal → Block. 30° → Block. 60° → Pass)

This says that if both polarizers are horizontal, the two photons do different things, which contradicts Observation 1. So the two photons in a pair must have the same hidden properties.

Let's consider again the first example of hidden properties:

(Horizontal → Pass. 30° → Block. 60° → Pass)

If all photons had exactly these properties, the photons in a pair would always do the same thing (Pass) whenever one polarizer was horizontal, and the other was 60°. But this should happen only 25% of the time, according to Observation 3. So a certain fraction of photons might have the hidden properties listed above, but other photons must have different hidden properties.

Now, let's list all possible hidden properties, in four groups. Two sets of hidden properties are in each group.

Group 1:   The two photons do the same thing, regardless of the orientation of the polarizers.
(Horizontal → Pass. 30° → Pass. 60° → Pass) and (Horizontal → Block. 30° → Block. 60° → Block)

Group 2:   The two photons do the same thing, unless exactly one of the polarizers is horizontal.
(Horizontal → Pass. 30° → Block. 60° → Block) and (Horizontal → Block. 30° → Pass. 60° → Pass)

Group 3:   The two photons do the same thing, unless exactly one of the polarizers is 30°.
(Horizontal → Pass. 30° → Block. 60° → Pass) and (Horizontal → Block. 30° → Pass. 60° → Block)

Group 4:   The two photons do the same thing, unless exactly one of the polarizers is 60°.
(Horizontal → Pass. 30° → Pass. 60° → Block) and (Horizontal → Block. 30° → Block. 60° → Pass)

Our final task is to determine the fraction of photon pairs that have hidden properties from each of the four groups. Let F_1 be the fraction (between 0 and 1) of photon pairs with hidden properties from Group 1. F_2, F_3, and F_4 are similarly defined.

Now, let's consider Observation 3, based on one horizontal polarizer and one 60° polarizer. The two photons do the same thing 25% of the time. Which of the hidden properties make the photons do the same thing for this combination of polarizer angles? Groups 1 and 3. This means that 25% of photons pairs must have properties from these groups:

$$F_1 + F_3 = 0.25$$

Next, let's consider Observation 2, in the specific case of one horizontal polarizer and one 30° polarizer. The two photons do the same thing 75% of the time. Which of the hidden properties make the photons do the same thing for this combination of polarizer angles? Groups 1 and 4. This means that 75% of photons pairs must have properties from these groups:

$$F_1 + F_4 = 0.75$$

We repeat for the other case of Observation 2, with one 30° polarizer and one 60° polarizer. The two photons do the same thing 75% of the time. Which of the hidden properties make the photons do the same thing for this combination of polarizer angles? Groups 1 and 2. Therefore:

$$F_1 + F_2 = 0.75$$

Finally, we use the fact that the sum of all fractions is one:

$$F_1 + F_2 + F_3 + F_4 = 1$$

If you subtract the first three equations from the last one, you find $F_1 = 0.375$. Plugging this into the first equation yields $F_3 = -0.125$. But we can't have a negative fraction of photons! Therefore our assumption of hidden properties is disproven! The photons do *not* have their observed properties prior to observation; prior to observation, the photons are in a mysterious indeterminate state.

Upside the Head

Marissa Lingen

20 February 2025

A professional hockey team funds research into concussion-induced brain damage, but the principal investigator worries about reaching her patients as people, not players.

I wasn't there to see it. I don't supervise the patients' lounge, mostly. Ben and the rest of the nurses and orderlies do that for me. I observe the patients in careful, clinical settings. I write up careful, clinical notes. All the useful information that isn't quite so careful and clinical comes to me through Ben and his team.

"Peter has a broken wrist and three broken teeth," Ben reported to me after it was sorted out. "Could've been worse. Nearly was. But Stosh was clearly winning, so the other boys didn't jump in on his side against the cop."

"What about Kendra?" She was the odd one out among our patients, the only one not tied to the Michigan Squids hockey franchise: a combat veteran, there so our sponsors looked good for Supporting Our Troops.

He shrugged. "Kendra's all in favor of law and order, but she knows when Peter's got himself into his own trouble."

"And he had."

© The Author 2017
M. Brotherton (ed.), *Science Fiction by Scientists*, Science and Fiction,
DOI 10.1007/978-3-319-41102-6_7

Ben eased himself into the chair across from my desk. I could see a bruise coming up under the dark skin of his chin, from where he'd taken an elbow or a head-butt, separating the two of them. "It was about the team. What they had or hadn't done to Peter's brother. Who loved Peter's brother, who supported him. Whether he was or wasn't a junkie."

"So…nothing left untouched."

"Basically."

"Peter's lucky the rest of the boys didn't pile on."

"Yep. Well. Ed might've added a choice comment or two. Ed might even have started the commentary. But I managed to keep him out of the actual brawl."

I shook my head ruefully. In his mid-sixties, Ed could afford to leave the fighting to the younger men, but that didn't mean he always had the sense to do it. He had never been able to keep his mouth shut. In his day, no one wore helmets, much less visors, so you could see him in every game — every minute of footage with him playing — jawing away at the other team, the ref, everybody. All the hits he took to the head had not added to his self-preservation.

Concussion research is like that. They are sweet. They are incredibly sweet. And then out of the blue, they will be triggered by nothing, or what looks like nothing, and there will be violence, or crying, or shouting, raised voices, or just…sullen silence.

Kendra is the worst for sullen silence.

Or there will be gaps, when they stare off into space, just nobody home. Stephane is the worst for that. He doesn't want to be part of our research at all. His wife made him.

I don't know what Ben saw on my face, but he said, "Give it time to work, Catherine." He calls me Catherine when no one else is around. It's all "Dr. Huang" in front of patients and staff, but we've worked together for decades now. "Rome wasn't burnt in a day."

"And amygdalas take time to grow, even in rats, much less in people. I know." I sighed. "I just — had hoped that they would be done beating the crap out of each other by now."

"'Patient socialization is one of the hallmarks of the Squids PCS Research Facility,'" Ben quoted from our brochure. I rolled my eyes.

Getting sponsorship from the Michigan Squids hockey team really turned a corner for us. It let us do all sorts of things we couldn't otherwise do — mostly it let us do everything faster. And God knew the Squids had provided us with plenty of test subjects over the years.

Stosh Majewski was the youngest, the most cheerful, the one who seemed the most unlikely to be in a medical research ward all the time — unless you saw his crying jags, his violent spells, his memory losses. Or unless you'd seen the video, which every hockey fan and half the rest of the internet had.

It was the video. The video where an opposing player drove his head into the ice from behind. After that video, it was hard to believe Stosh was alive, much less walking around, talking, joking with his doctors. Hockey fans had explained to me that it was Todd Bertuzzi's hit on Steve Moore with the volume cranked up to eleven. I didn't know what that meant. I just knew that Stosh was only twenty-four, and he wanted his balance back. He wanted his memory back. He wanted back in the game.

And his old team, the Squids, had faintly dawning hopes that my amygdala regrowth factor would be the miracle drug that could make that happen.

Amygdala regrowth factor. ARF. Stosh had the other patients barking like dogs when Ben and the other nurses came at them with injections. We needed a better name before we went public with it.

First we needed results.

First we needed the patients to grow enough amygdala back that my best nurse wasn't beaten all to hell from pulling them off each other.

"Put some ice on that bruise," I said to Ben. "And if Stosh brings up Jesse again, get him out of the room, fast. Or get Peter out of there."

"I'll see what I can do."

Peter's brother Jesse was the reason he was in there. The Squids were trying to avoid a lawsuit from the grieving Van den Berg family. Peter had only gotten as far as the minor leagues, then retired to be a cop after one too many hard hits. His brother Jesse was a fan favorite, a gentle giant with the fans and the last of the great enforcers on the ice.

He'd died in agony of a drug overdose. None of the fans had known until it was too late. He'd come from a cop family. It seemed so unlikely. But the pain from the fights, the brain damage from all the concussions — his judgment was shredded. And the Squids' team doctors…let's say that one of my conditions when they hired me was that I didn't have to meet with the team doctors. I wasn't going to shake any of those guys' hands.

Ben shook his head. "Give it time, Catherine," he repeated. "It's a long game."

He left my office. I looked at the rest of the files. I couldn't see anything yet. Maybe there wouldn't be anything *to* see.

They all had hope. I would have to.

24 February 2025

Stephane's wife came to visit today while I was having a session with him. Claire. Her English is so great. She's never complained about having to move to Michigan for the treatments, even though I know it was hard for her to find work, even though she knows no one but the other Squids' wives.

"You're making progress, I can tell," she said.

Stephane looked away dismissively. That's all he ever does. I felt like I had to look away, too; I left the room to give them privacy. I know that one of Stephane's main symptoms is depression, but in this case he's right: his memory loss isn't showing any improvement so far. Maybe that means he's one of the placebo patients. Maybe the ARF just doesn't work on humans, or maybe it can't undo the damage from old trauma as much as we'd hope — just stop things from getting worse.

I went down the hall to look in on the patient lounge. They had a hockey game on. I've tried to learn the game better since I started working in a Squids-funded facility, but to tell the truth I can never recognize offsides, never.

I almost walked past, and then I heard Ed and Stosh reminiscing. They sounded like old drinking buddies — not arguing, just shooting the breeze.

"I haven't seen a hat trick that pretty since Bruno Detwiler in '22," said Ed.

"March of '22, yeah," said Stosh. "That was my second year. God, that was a great hat trick. No assist on the third goal, either, and not an empty netter."

I froze. Only three years ago. One of the hallmarks of their kinds of brain injury is difficulty in forming and accessing memories from recent years. Was I reading too much into this? I went and wrote it in my notes anyway, just in case.

"Put a sock in it, boys, we're trying to watch the game," said Peter good-naturedly, and no one jumped on him.

20 March 2025

I got a call from the Squids organization today, what they insist on calling "the front desk." (To me the front desk is my secretary, Wescott.) "Great news!" said the suit, whose name is Bill. "Looks like we're for sure going to make the playoffs this year. Got a pretty good chance at the whole thing!"

"That's great!" I said. I think I even managed to sound sincere, too. It helped that Ed had performed better on one of the memory tests than he ever had, so I was in a genuinely good mood. Ready to be happy about other people's triumphs.

"We're going to want a news spot about your research for the period breaks," he said, like it was the most natural thing in the world.

"I…beg your pardon?"

"A feature. Just a puff piece, one of those misty feel-good journalism thingies," he said. "Our press gal will set it up with you."

"My patients are in a very delicate state, Bill," I said. "It's the early stages of their amygdala regrowth. I really don't want them disturbed."

"I understand," he said. "It'll all be very gentle, very soft focus. No hard-hitting questions. It's not like you're going in front of the Inquisition, ha ha ha."

"Ha ha ha," I said dutifully.

"But we're not in this for our health, you know. We want to get some good PR out of it."

I knew what I was doing when I left Michigan State. Or I told myself I did. "PR. Of course."

"Great! I'll have her set it up with you. I'm sure it'll tug everybody's heart-strings, you'll be fabulous. You always look so serious and scientific."

I wasted probably five minutes after terminating the call trying to figure out whether that was a racial remark. I decided I didn't care. I called Ben into my office.

"Brace yourself, the TV cameras are coming in."

What Ben said should probably not be recorded for posterity, but we've known each other long enough that he knew I felt the same.

He did, however, tell me that when one of the orderlies ran Kendra's foot over with a cart, Kendra not only didn't punch her, she said, softly, "Hey, hon, it's okay. We all make mistakes."

And then went to talk to Ed about Ed's kids and their vacation to Florida, what she had liked best about her own Florida vacation before she was deployed.

Is that…apathy lifting? Empathy, control of temper, management of memory? All at once?

Or is it just a good day?

Good days don't count until they add up. Write it down. Write it all down.

22 March 2025

Claire has heard about the TV spot. She came into my office worried. "They're not going to mention Stephane's drinking, are they?"

"I don't think so," I said.

"There was — " She hesitated. "The DWI. It was a matter of public record. I don't think we can hide it, Dr. Huang."

"This is not investigative journalism, Claire." I knew I was supposed to do something to make a connection, so I reached across my desk and patted her hand awkwardly. "The team is doing this. They want everybody to feel good about the players, about the research they're funding. That means showing health, recovery. Not darkness and mistakes."

She looked unsure.

"They'll probably mention — in general — that one of the symptoms that *some people* have is difficulty handling drugs and alcohol. And they may focus on Jesse, with that. But I'll speak to the producer, make sure they know of your concerns."

Claire's hand flew to her mouth. "Jesse. I hadn't even thought of Peter and the rest of the family."

"I will — " I wanted to tell her I would handle it. I still want to say I can handle it. But they have the final word on funding. They're going to do what they'll do. "I will let them know."

Speaking of the final word, the front desk — not Wescott, the other one — has let me know that ARF has been renamed cephalladine in time for the video shoot. This is supposed to sound vaguely reminiscent of squid, for the Squids fans who are smart enough to know root words, and just generally scientific for the rest.

I have told the patients. They are still barking at it. Ed tried making wiggly squid fingers, but no one else picked that up. Ben told them if they barked on camera, it would get cut out of the footage and also he would tell the kitchen that they didn't like peanut butter pie for desserts any more.

Much grumbling. Less barking.

29 March 2025

It could have been much more catastrophic, this video shoot. There were some volume control issues. Even with the ARF — excuse me, cephalladine — some of them just don't have any idea of how loud they're talking. Even when they're not agitated. Maybe that'll get turned down in production. Maybe it's just the sort of thing they want.

I'll be excited to see the results when we can look at the "after" MRIs without compromising the treatment. I have so many theories about who is and isn't in the control group. But "double blind" is important.

We got the whole group together to shout, "Go Squids!" for the end. Promotional material. This is my life now. It pays the bills, I guess.

I never expected Stosh Majewski to be the one to roll his eyes at me as we all wandered off our separate ways, though.

23 April 2025

The Squids got us seats to one of their playoff games, very good seats, not rinkside but very good. We sat in a group.

The TV journalist did not interview me or Ben or any of the other staff. We are in the PR spot, being earnestly interviewed and scientific. We know our place. In the timeouts, in the period breaks, they want a few minutes with the old players.

And that is where it happened.

Ed was being interviewed live on national TV — international TV, of course it was across the border not just in Windsor but all over Canada. There was some caption under his name, "Ed McCann, Squids Stanley Cup Team '90, '91, '93," something like that.

The guy tossed him an easy question — "Isn't it amazing to see the Squids in the playoffs again after all these years?"

And Ed said, "It is, Dave — " The guy's name really was Dave. Only six months before Ben and I would have high-fived each other that he could remember Dave was Dave and not Jerry or Braden or Junior.

"It is, Dave," said Ed, "but what I'm really concerned about right now is the escalating situation with the water wars in Bangladesh. With all the energy we're pouring into the ice tonight, I think *that's* the water we should all be concerned about."

Dave — not Jerry or Braden or Junior — had to cut back to the color commentator and the play-by-play announcer, going back to the main part of the game, because the faceoff was starting anyway.

So we didn't get in too much trouble.

And I don't think they'd yank our funding over Ed McCann running his mouth on TV, because this is the guy who referred to the Canadian Prime Minister as "that frigid shit-for-brains Canucks fan" on a live feed, right before he got put in our program.

But Ed. Our Ed. Being able to pronounce Bangladesh, much less knowing that there's a problem with the water supply there?

I'd say that's neural regrowth, there. In areas none of us could anticipate.

I wrote it down in my clinical notes as "personality changes."

But in this private notebook I can just say: go, Ed.

And please shut your mouth and say, "Go Squids" for the rest of the play-offs, for all of our sakes.

30 April 2025

Ed trashed his room today. Maybe I was wrong about what the Bangladeshi water thing meant, but…I don't think so.

The sheets were shredded. The bedframe broken. He had a plant in there that he loved. The pot was shattered, the root ball in pieces.

Ben watched to make sure he didn't injure himself, but otherwise he just closed the door and left him to it. Safest thing. A guy like Ed, he's a really nice guy in his best moods, but he knows how to destroy things. I've advised the staff that they don't have to make it *them*. Property is just property.

The Squids can afford new bed frames, after all.

When I asked Ed about it later, when he calmed down, all he would say was, "Man, Doc, I don't know. I don't even know who I am these days."

Well. I think that would upset anybody. Especially anybody with decades of never learning any other coping skills.

I've made notes about having occupational therapy to teach other coping skills in future trials.

I feel like a cold and terrible person that I think those therapy classes would have tainted these trials.

Especially when I know that I'm sending half of these people home to their families without either the cephalladine effects *or* the therapy.

I tell myself it's for the longer-term good of everyone with this kind of brain injury. I hope I'm right. I watch Stephane staring at the corner of the room, not even the TV, just the corner, and I think of what Claire will face when I give him back to her. He didn't want into the trial in the first place — she was the one with hope. Couldn't I have given her husband a couple of emotional training wheels?

Apparently not.

12 May 2025

The Squids' drive for the Stanley Cup is still going strong, but they've lost the last fan I ever thought they'd lose.

Stosh wandered away from the patient lounge as I was packing up to leave for the night.

"I don't know, I just couldn't stand another night of hockey on the TV," he confessed. "I think I'll just go for a walk in the garden."

"You feeling all right, kiddo?" I asked. I resisted the urge to take his temperature. Ben does that every day.

"Sure," he said. "I've just had enough hockey. Weird, huh?"

"You could read a book or something," I suggested.

He laughed. "Come on, Doc, let's not get crazy about this. It's pretty nice out. I'll just stretch my legs. Probably listen to music on my device."

"Sounds nice."

Despite his lack of interest in reading, I wrote, "Personality changes," in my chart next to his name: I realized that while his balance has improved, he hasn't asked about getting back on his skates in months.

Ed is in watching the game with the others, but he's also filmed an appeal for the water charity in Bangladesh that approached him after he opened his big mouth in April.

I'd have liked to see some personality changes in Stephane — Claire is sure she sees them, but I can't agree with her. He still mostly sits and stares. And Peter is as sullen and defensive as ever about Jesse's memory, but who knows whether that's concussion-induced irritability or a sincerely held belief. Maybe some of both. Put my brother through what Jesse went through, and I'd probably be pretty snappish about it, too, regardless of what state my amygdala was in. The memory and coordination tests will be a better gauge of what's going on with Peter.

I wish I could say with Kendra. I wish I could say *anything* about Kendra. I think this was the wrong program for her. Being the poster child for Supporting Our Troops is not going to help anybody with their depression and isolation — especially when you're the only one who has no connection to the thing everyone else is connected to.

She's in there watching the game with the others, though, even asking polite questions, so who knows.

The front desk has decided to content itself with a picture of all of us together in Squids playoffs promotional gear. The front desk is not as dumb as it looks.

28 May 2025

The front desk came into my office personally to teach me the facts of life.

"Stosh Majewski is a young man with his life ahead of him," said Bill. His suit cost more than our portable MRIs. The researchers of yesterday — the developing nations of today — would cure hundreds of kids for what his shoes cost.

"The more so now," I said through a tight smile.

Bill's smile didn't falter. He is better at this than I am. "We are not paying for Stosh to have therapy because we like him."

"You should. He's a really nice kid."

"He's a really nice kid with a hell of a one-timer."

"I don't even know what that means."

Bill sighed gustily. The fern on my desk rustled in his minty fresh breeze. I wanted to cup a protective hand around it. Irrational. "Dr. Huang — Catherine — we pay your salary. We keep you and the players — "

"Patients."

" — patients," he agreed smoothly, "in this nice facility. But Majewski has millions left on his contract. If he *can't* skate again, we're not monsters. What we're hearing is that he *won't* skate again."

"For me, that's a distinction without a difference."

"For us, the difference is several million."

I waited. But he is, as I said, better at this than I am.

"Catherine," he said, emphasizing that this time he was not even trying on the use of my proper title. "You're going to have to find out for sure whether Majewski is actually still sick and unable to play or just malingering."

"That's my job?"

"We think it is."

"I think my job is to find out whether I can heal these patients and report back to the world — and to you in particular — about a very serious problem that plagues your industry."

"Of course! No one said not to."

"And get you Stosh Majewski back in a Squids shirt."

"*Sweater*," he said, and finally the smile wavered.

I didn't tell Stosh about this conversation. I didn't want to upset him. Instead I told Ed. He looked so fierce I thought maybe I'd made the wrong choice, but after a moment he nodded, hard. "Don't worry, Doc. We'll get the boys together. The team needs to know we've got Stosh's back."

I couldn't see how that was enough, but Ed kept nodding. "You understand a lot about the brain, Doc, but you don't know the part that has teams in it yet. We'll get this fixed."

29 May 2025

Ed must have been right. It was Stephane who went to the front office. Stephane.

There wasn't even any yelling.

But they've eased off on Stosh, and our funding is approved for another several rounds of testing, not even with former players for the main body of subjects.

For as long as I've been working with these guys, I don't think I'll ever really understand them.

Stosh is going to be in three commercials and on five new billboards, though. Stephane thought of that, too, with Ed's help.

And then he came home and punched an orderly.

I swear this job is going to give *me* difficulty with emotional response, volume control, and appropriate regulation of alcohol consumption.

10 June 2025

The Squids are in the final round of the Stanley Cup playoffs. Even Stosh is watching for old times' sake.

We haven't had any fights since March. Grousing, arguments, sure. But no fights. Half these people are on placebo. But I think it's clear that we're seeing less irritability, which probably means some amygdala regrowth, some brain healing — only half as many of the people in the patient lounge are dealing with these problems, which has to reduce conflict all by itself.

Also maybe they're used to each other. Who knows.

Claire came into my office and cried today. "Stephane is on the placebo, isn't he. He's not really improving at all."

"I'm not allowed to know that until the trial is over," I said, as gently as I could.

"I've been fooling myself."

I thought hard about what the best thing to say was. "Maybe you've just been encouraging him."

"I want my husband back!" she sobbed. "I want Stephane!" She wore lots of gracefully applied eye makeup, and it was all clumping and running in the least graceful possible manner. I handed her tissues as fast as I could pluck them from the box. I made soothing noises.

I am not a clinician. I am not good at soothing noises.

Finally, when she started to calm herself, I ventured something less soothing.

"If he is receiving the placebo, that's good news," I said.

She sniffled and looked inquisitive.

"If he is, he can perhaps receive the real thing, when it goes through to the next stage, or is approved. *Perhaps*," I emphasized.

Claire looked more hopeful. "It's working on some of the others."

"I think so. But I have to warn you. There are personality changes."

Alarm. "Not — some of the concussion men, they are — more violent — "

"No, no," I hastened to reassure her. "Quite the opposite, or — not the opposite exactly, just — it's hard to predict. They're going off in directions we can't anticipate. You saw Ed McCann talking about Bangladesh on TV perhaps."

She waved a well-manicured hand. "Oh, well, Ed."

I stifled a laugh, turned it into a cough. Perhaps I am overstating the personality changes. Perhaps Ed will do what Ed will do.

The personality changes would be harder to live with if she was actually living with them. I knew that much. Finding that someone's decisions and emotional memory were restored sounded grand, but they weren't restored, they weren't recreated, they were — made into different ones. And what are we if not our decisions and our emotional memory?

But I looked in on her and Stephane together, and I thought that perhaps with what they had been through she would shape her own memories to him, if she could have him well again. She would tell herself that this was how he was.

The true test will be releasing my patients into the world. The casual friends and acquaintances, the more distant relations who can't visit every few days — will they feel that they still know them? Will they find anything in common, still? Or will those memories be reshaped, changed so that "of course" Ed will be Ed, obviously he was always like that?

I used to think that going into the world would test their self-control and inhibitive behavior for drug and alcohol use, and perhaps it will. But as I grow more confident in the regrowth factor, the rest of the data — which is much harder to gather — interests me far more.

Afterword

Head injuries are big business these days. We're spending a lot of time trying to figure out how to prevent them, and that's a really good thing — an ounce of prevention, as the old saying goes. But sometimes people *don't* manage to prevent concussion. Sometimes — as hard as it can be for those of us who make a living with our brains to understand — it's not their top priority.

One of the main things that goes wrong with multiple concussions is amygdala damage. The amygdala — properly speaking, the amygdalae, you've got one per side — is in the middle of your temporal lobe, and it's good for all sorts of useful things: memory, decision-making, and how you process your emotions. Combined with cultural factors, this is why athletes (and soldiers and other people with repeated head trauma) end up with a reputation for anger management issues: they have literally broken the parts of their brains that are meant to handle that.

You can read about post-concussion syndrome and amygdala damage at whatever level you like. The Mayo Clinic website has a brief overview of post-concussion syndrome at http://www.mayoclinic.org/diseases-conditions/post-concussion-syndrome/basics/definition/con-20032705. Almost any major metro newspaper will carry stories on football players and post-concussion syndrome in their archives of the past two to three years. On the more technical side, academic journals are filled with papers like "Relationship between competency in activities, injury severity and post-concussion symptoms after traumatic brain injury," by Unni Sveen, Erik Bautz-Holter, Leiv Sandvik, Kristin Alvsåker, and Cecilie Røe, or "Long-term structural changes after mTBI, and their relation to post-concussion symptoms," by Philip J. A. Dean, Joao Ricardo Sato, Gilson Vieira, Adam McNamara, and Annette Sterr. ("mTBI" means "mild traumatic brain injury." It means your brains have been jostled nastily but are not leaking out your ears.) You can easily go down the rabbit hole of reading as much on that as you want or as little.

The politics of whether people *should* go around knocking each other in the noggin might be simpler if someone could just reach in and fix an amygdala like a broken ankle. Would that be better? I don't know — neither does my main character.

On a more personal note, I have a non-concussion related balance problem. I decided not to get into the balance problems of these patients, as their emotions were more interesting to me, but the vertigo that can come from head injury is not any more fun than the kind I suffer from, the kind that comes out of the blue. Hockey is also one of the more fun ways to get head injuries. Be safe out there, and keep each other safe when you can.

Betelgeuse

J. Craig Wheeler

Fire fills the sky from horizon to horizon, hot red and yellow, surging, billowing, flaring gas. I'm in close orbit with flaming Betelgeuse horizon to horizon. Very impressive. In the opposite direction, overhead, is cold, dark sky.

Being out here gives you a different perspective. I've had a long time to think about this. I'm not sure it's a good idea after all.

Things evolve. Even the notion of what it means to think changes. I had my active part in the decision. The science was compelling. We've learned so much about the Universe, but there is so much more to learn. That drive to understand doesn't go away.

My quest is a microcosm of all the exploration going on, a chance to add some details to the overall scheme — a cloud of gas, a contraction, a spinning up, a breaking apart, two stars forming, one swells, they merge, a doomed core, a collapse, a violent disruption, seeds of new beginnings. We've known all that for a very long time, but the information came from remote reaches. This was the chance to check out the process right in the neighborhood. Cross some i's, dot some t's. Who knows, maybe the lovely goddess Serendipity would grace us with tantalizing unexpectedness.

From Earth, Betelgeuse had hung in the sky, resolute, during the Transformation. Some things can't be hurried; they don't respond to the hurly burly, the comings and goings of living things. Or they just live their lives at their own sedate pace. Until the end.

We'd been monitoring Betelgeuse. We knew the clues to seek. Amid the long, slow tumble of the outer envelope, there was the faint, high-frequency noise of the inner chatter. Burning shells heaved and surged, sending small,

© The Author 2017
M. Brotherton (ed.), *Science Fiction by Scientists*, Science and Fiction,
DOI 10.1007/978-3-319-41102-6_8

but unmistakable signals, writing the evidence of their condition first in seismic waves and then on the surface light racing out to instruments on Earth. The timescales got shorter and shorter, signaling the encroaching end. The icing was the faint but detectable rising signal of thermal neutrinos from the detectors floating in the quiet oceans of Europa.

It seemed logical. Off I went.

—//—

"Protocol confirmed. Acknowledge, 7832299081."

"Hello, 43850388443. What'cha up to?"

"I was up to 0.99 light speed. I'm into deceleration now, poking along. I just got out of hibernation."

"You're the Betelgeuse core mission, right?"

"That's me. Records show you're a life-seeker."

"Not much to show for it. I've probed hundreds of planets. Weeds, lichen, some bugs. No sentient life. Nice to hear a voice."

"Me, too. I picked up your solitary signal out of the noise as home was fading."

"Yes. I've been tracking you for awhile."

"I'll be at Betelgeuse shortly. You're nearby?"

"Yes. Just wrapping up here. I should be in the Betelgeuse system by the time you make orbit.

"It will be easier to talk then. I look forward to it."

"Me too. Take care passing the ice line. Apparently a lot of dark nasty stuff still hanging around out there."

"Roger. Until then."

—//—

"Protocol confirmed. Acknowledge, 43850388443."

"Well, hello, 7832299081. How've you been?"

"Welcome to Betelgeuse! I'm fine. Got here 20 sols ago and set up shop. I've been getting some data on the outer planets. Any trouble getting into orbit?"

"Nope, smooth as silk. Looks as if we are going to be here a while, just you and me. Can I call you '7?'"

"Whoa, we just met! We should be more formal for the time being. You can call me 78."

"That'll work, nice to meet you, 78."

"Same to you, 43. What do you see?"

"Fire and ice. From this close orbit Betelgeuse fills half the sky. Quite a sight, the outer convective envelope. Then the other half of the sky is deep space. Nearby is mostly dark, but that depends on the band. Betelgeuse blows a pretty healthy wind, and that shows up strongly in some frequencies. How about you?"

"I'm out far enough that Betelgeuse only spans about ten degrees in my field of view. Even here it's quite a sight. Boiling cauldron. You're right about the wind, 43. There are some terrific aurorae on planets and moons that have magnetic fields, large distortion of magnetospheres, that sort of thing. Ah, I see you now, 43, little spot of a shadow."

"Can you see me waving, 78? What have you learned?"

"It's been a maelstrom around here for a long time. Betelgeuse used to be a binary system."

"I trained for this mission, 78. I have memory banks. I know the history."

"Don't be snotty, 43. You asked, so listen and learn. Twin stars are already hard on planet orbits, so there is a lot of chaos in that regard. Then about a hundred thousand years ago, Betelgeuse swelled up to become a red supergiant and coalesced with its companion…"

"Right, made that expanding bow shock, point four parsecs out."

"… causing even more irregularity in the local gravity field. Some of the planets out here formed when the Betelgeuse system originally did. Those used to be in the outer cold depths of the system before Betelgeuse expanded. Some were flung outward during the merger, some formed in the wake of the merger. The very oldest planets are only eleven million years old, too young for life to have formed. They're all barren."

"Quite a history, 78."

"Never mind the inner planets that were roasted when Betelgeuse expanded. I guess the future is not so bright for the rest of them, either. Or, too bright. Pun intended."

"Funny, 78. Like your work?"

"I do. In the first place, it's gorgeous out here, the bright vivid specks of the stars. Then you get up close to the planets, and they're all different, even the bare rocky ones, but especially the ones with atmospheres, water, clouds. Like the Earth long ago. In the data bases. Takes one back, even if one didn't live then."

"Ah, a romantic."

"Guilty, but the science is also fascinating, 43. It's a great kick when the biomarkers pay off, signs of living things, even if it's just cryptoendolithopsychrophiles. There are common themes, but life finds different ways to twist the coding molecules."

"Life finds a way."

"That it does, 43."

"But no sentient beings, you said before."

"Not yet. It's the ancient debate. Are we the first? Are they there, but their technology is so advanced it's magic?"

"Hyperspace would be like that, magic. Why use modulated lasers when you can chat instantaneously in 4D? It's frustrating to know hyperspace is there, but not be able to get to it."

"Patience, 43. Hard work is being done on that."

"Failing that, it would be nice to have a knob, just dial the speed of light up when you want it."

"Spoken like an engineer."

"I guess."

"You're just like so many others, 43. You want to solve every problem. Can't you just appreciate what we've got?"

"I appreciate, 78. You can't help dreaming for more. Besides, we're wired to solve problems."

"I suppose."

"Speaking of that. Back to work. Talk to you later, 78."

"Deal."

—//—

"Protocol confirmed. Acknowledge, 7832299081."

"How are you doing, 43?"

"They don't quite prepare us for the solitude, 78, do they?"

"No. You can study it, learn about it, but the experience is different."

"Billions of conversations at home, rapidly fading to very few, often none on the trip out. Took my mind a while to adjust. I felt as if my brain were about to explode, out into the vacuum. I still get billions of faint signals, but all one way, essentially no conversation, no feedback. Except for you."

"Similar for me, 43. I was already pretty isolated from the home planets, before I started to pick up your comms."

"What do you think about, 78, out here by yourself, without all the interaction?"

"Besides my mission, you mean?"

"Yeah. It's so weird to be alone with one's thoughts for so long."

"I spend a lot of time in the memory banks, browsing. Everything everyone has ever known or done, is right there. That's always true, but at home you don't have time amid all the distractions. Out here, there is time."

"… It's so damn quiet."

"That it is. I guess this is what it was like for the first million years of human existence. 43 grunts, 78 grunts back. Before everything got interconnected."

"Kinda atavistic."

"There are billions of us even in our small part of the Galaxy, but most are 640 years away from us here and now. Conversation is a tad slow under these circumstances."

"Even thinking is different, 78. Back home, thinking meant folding multiple conversations at once into one's thoughts. You're thinking others' thoughts, and they yours. Here it's just yourself. And my life-seeking companion, 78. There were lyrics in an old song, 'Don't let the sound of your own wheels drive you crazy.'"

"You're not losing it, are you, 43?"

"Nah, it just takes some getting used to. Do you ever think about where we've come from, where we're going?"

"Sometimes."

"Being out here makes you feel kind of small."

"Actually, 43, this experience makes me feel connected, to the whole Universe, to all that's going on. To the growing complexity. The rate of increase of knowledge proportional to the base of knowledge, the recipe for exponential growth. Part of all that."

"Even though we're expanding to a dilute heat death."

"That's a long way off. Lots will happen."

"I'm used to being we. Out here it's me. And you. Do you think about that, 78? What makes you, you?"

"Nature and nurture. My roots, and yours, go a long way back, to when the first molecule folded, bonded to itself, and became self-reproducing. Cell division then sex. Our nature goes to those roots, but that nature has gotten ever more complex, preserving some history, neglecting other aspects. We have common roots, but that does not keep us from being different, individuals, me and you. Then there is experience. We've been different places, seen different things."

"All that's in the data banks, 78. We share all that knowledge, that history."

"In principle, but you can't access it all, all the time. We access it differently."

"So, we're different."

"That we are, 43."

"Takes some pondering. I'll get back to you, 78"

"Until then."

—//—

"Protocol confirmed. Acknowledge, 7832299081."

"Hi, 43."

"Uh, 78, uh…"

"Spit it out, 43. What's on your mind."

"I like you."

"..."

"This is not a familiar feeling to me, 78."

"Good or bad?"

"Uh, good?"

"I bet you say that to all the life-seekers."

"Back home, it's not even a concept. You have to get to know someone. As an individual. As we have."

"'Like.' Do you even know what the word means, 43?"

"I'm learning. I like that you're different. You see things differently."

"Yes, I'm further away from the star."

"That's not what I mean, and you know it, 78."

"I do know what you mean, 43. I like you, too."

"… This is awkward. I'll talk to you later."

"Bye, 43."

—//—

"Protocol confirmed. Acknowledge, 43850388443."

"Hey, 78. What's up?"

"You won't believe this, 43."

"Try me."

"I found a planet with life on it, advanced life."

"You said they were all barren."

"This one is old. It can't have been born in the Betelgeuse system. It's a rogue planet, born elsewhere, apparently drifted in, captured by the local gravity."

"Left its host solar system? What, heated by internal radioactive decay?"

"Yes! Heated from within, a dense atmosphere traps the heat. It's essentially covered by a warm ocean."

"There is a record of such things. Small probability, but finite."

"But I haven't found one before! And here, around Betelgeuse."

"That is pretty remarkable. Congratulations, 78. So. What? Bacteria? Stromatolites?"

"No, you don't understand, 43. Animals! Aquatic animals! It's difficult to do a census, but millions of them, maybe billions. And they communicate! I'm gathering data to decipher, but no question. I think they build things."

"Build things?"

"Yes! There are shapes on the ocean floor. I can make them out in the shallower waters. I don't know their function, but they show rectilinear patterns. I'm sure they're not natural."

"You're going to be famous, 78."

"That's not the point. The point is, they are going to die!"

"Life and death. That's how things work."

"I don't mean that, individuals. I mean extermination! All of them, the whole planet. When Betelgeuse goes up."

"Inevitable, and soon."

"Don't be so hard hearted, 43! This is the annihilation of a species we are talking about. Maybe an advanced, conscious intelligence."

"I appreciate that, 78. It's tragic, but there's nothing to be done. This star is going up."

"I hate that!"

"It's disappointing. But if there's one, there will be others. In the meantime, you need to collect all the data you can."

"But I'm alone! Can we bring others? There's so much to be done."

"No. No time. You'll have to do what you can."

—//—

"Protocol confirmed. Acknowledge, 7832299081."

"43?"

"Its time, 78. Nice sharing these last several years with you."

"I've been thinking, 43. You don't have to do this."

"What do you mean? It's what I'm here for."

"That doesn't mean you have to do it."

"Of course I do. It's my mission."

"You can fire your boosters. Come out here. Join me, 43. Share the demise of my planet with me. Then we can take off. See what happens. See what we find."

"That's nuts! I have data to collect. This is a once in a million-year opportunity."

"My planet is going to die. Your mission could die. You don't have to."

"I've loved this time with you, 78. I had no idea, but I have a commitment."

"Your only commitment is to yourself. And to me."

"… I've got data modules to prepare."

"43…"

—//—

Here I am.

I've spent these few years here, probing Betelgeuse. I got here just after the carbon core lit. I watched the oxygen come and go, the silicon core form a few days ago when I last exchanged with 78. Then the iron, now minutes from its limit. Then all hell breaks loose.

The plan is, I hit the boosters and dive inward at the instant of collapse. I'll leave behind a string of data modules. They are tiny and tough. Packed with memory. A lot of them will survive. They'll be riding out with the ejecta or orbiting the pulsar thousands of years from now when the follow-up crews arrive.

I'll meet the shock wave in a few hours, just outside the helium core. I'll get as far as I can. I won't make it to the heart of the neutron star, but my physical limits will be part of the data to be collected.

As I said, all this made a lot of sense, back when I was a small tangled piece of the hive mind. Collect a little data, all for the advancement of knowledge. Being out here gives you a different perspective. Alone. Time to think. Getting to know another being. 78.

When I started this project I had no idea what "someone" meant. It was just an amorphous "them." Amorphous me, for that matter. The notion that you could be attracted to someone was totally, not just foreign, but inconceivable.

I can feel my metallic skin beginning to warm.

Afterword

Betelgeuse has been my obsession for many years. Betelgeuse is a massive star, fifteen to twenty times the mass of the Sun. We know enough to be confident that it will go though a series of nuclear burning stages turning hydrogen first into helium and then into ever heavier elements until a core of iron builds up in the center. From its nuclear properties, ordinary old makes-rust iron is *endothermic*, it absorbs energy from the star and produces none. The result is that after millions of years of evolving, the iron core in Betelgeuse will form and linger for perhaps a day, but then will absorb energy and trigger its own collapse. The collapsing iron core will produce a burst of ephemeral particles called neutrinos and will in the process form a neutron star, likely a rotating, magnetic pulsar, and a gigantic explosion that will blast the matter beyond the iron core out into the surrounding space in a brilliant supernova explosion. Thence are the elements to make planets, life, and people.

The roots of my obsession go back to when I more regularly gave popular talks. Someone would ask, "what happens when Betelgeuse explodes?" and I would say, "you know, someone asked me that the last time I gave a talk like this. I promised to look into it, but did not get around to it. Next time, I will." Then, of course, I would not, and I would go through a similar exchange after the next talk. This happened several times, until I finally took the time to do a little thinking. I wrote up the results as a sidebar in the popular-level book, *Cosmic Catastrophes*, I wrote for and use in my classes at The University of Texas at Austin. Betelgeuse will be very dramatic when it explodes, but it is far enough away, best guess being a little over 600 light years, that it will not be dangerous. It will be a single, intense point of light, about as bright as a quarter moon for about three months before fading. This effort got me a mention in the Wikipedia article on Betelgeuse.

This got me to pondering how long it would be until Betelgeuse explodes and to the realization that no one really knew. In this spirit, and only slightly tongue in cheek, I began to ask my students, non-major classes of about 200 apiece, to make Betelgeuse part of their "sky watch" extra credit projects. I would tell them that we know it will explode, and how, but not when, and that we are so ignorant that it might blow up tonight. I said, "if it starts to get really bright, let me know!" I say on the final day of class that if they remember nothing else from my class, I hope they take their grandkids out, point out Betelgeuse and say "some day that star is going to blow up!" All told, I left variations of that message with perhaps 3000 students over the years, maybe more.

After some time of this pedagogic exercise, I began to contemplate how deeply frustrating it is scientifically not to know when Betelgeuse is going to blow up. That was the beginning of what I call *The Betelgeuse Project*, an effort to determine when. My original notion was that to determine when Betelgeuse will explode, we need to know its evolutionary state and for that, we need to peer inside. The solution it seemed, and seems, to me was to use asteroseismology. That is the technique of studying light variations on the surface of a star to probe oscillations its depth. This is closely analogous to using the propagation of earthquake waves to determine the interior of the Earth with its outer crust and molten core. This has been done to great effect with the Sun, so that we know it has peculiar rotation properties, with its outer layers rotating on cones, rather than cylinders, as standard physics would seem to dictate. It has also been a powerful technique to study white dwarfs. Data from the COROT and Kepler satellites and other facilities have brought a wealth of data on small sensitive variations of surface luminosities that do probe the interiors of stars. People have been able to discern that certain red giant stars are burning hydrogen in a shell, but not helium and that others are also burning helium in their centers. Other people have been able to discern the effects of a magnetic field dragging on the rotation of a red giant core. There is a revolution going on in the study of stars with this new ability to peer inside.

Why not, then, I thought, apply this to Betelgeuse? My next notion was that what one would seek to detect would be the "noise" associated with deep convective burning shells. This convection represents a "boiling" that could make waves deep within the star that would contain clues to what was going on. In particular, just before the iron core collapses to trigger the explosion, the burning of surrounding oxygen and silicon layers are expected to be especially chaotic and vigorous. If one could detect evidence of that phase, you would know that the explosion was weeks or even only days away. I draw on this notion in the story.

There is an issue that Betelgeuse is surrounded by a large, boiling, hydrogen envelope making it a red supergiant star. One has to be able to "see" through that outer gas bag. The trick, I hoped, was that the time scales from the inner burning would be so short, days, that the outer envelope with motions on the order of years would be basically standing still. Perhaps, amid the long slow variation of the outer layer, one might detect a quick, faint, rattle that would indicate the state of the interior evolution and hence the time until explosion.

I went through a phase when I thought perhaps it is better to look at stars that will explode, but which do not have the confounding outer envelope. Those stars exist; they are known as Wolf-Rayet stars. I did some preliminary thinking about that, but realized the probability of catching any star, Betelgeuse or a Wolf-Rayet star, in the short time just before the explosion is statistically unlikely. I still think one should do asteroseismology of Wolf-Rayet stars.

Meanwhile, back at Betelgeuse, I was lucky to recruit a group of bright, hard working undergraduates to *The Betelgeuse Project*. Over the last several years, I have had the privilege of working with a young man from Physics who is now studying in Hawaii, a young woman whose mother is Egyptian, a young man from the Rio Grande Valley near the Texas/Mexico border, and summer students from China and Greece. We have been using the stellar evolution code, *MESA*, which is itself a revolution, an open source code with vast capacity and an active helping community. We have used *MESA* to compute models of Betelgeuse, including estimates of the characteristic inner convective noise that might be a clue to the internal state of evolution.

Our biggest surprise, however, came when we tried to match the rotational velocity. We could not. There was no combination of internal processes that would leave the outside we can see spinning as rapidly as it is observed to do. Some pondering brought a possible solution. We suggest that Betelgeuse once had a binary companion that was close enough in orbit that it was engulfed when Betelgeuse became a red supergiant. That orbital velocity might very well have provided the angular momentum to spin up the outer envelope. We also speculate that this merger might have led to some of the shells that surround Betelgeuse at a substantial distance. All this, too, is woven into the background of the story.

Sticks and Stones

Stephanie Osborn

:::LIVE DOWNLINK COMMENCING:::
International Space Station
Increment 58
Mission Elapsed Time 8 days, 9 hours, 47 minutes
Subject: ESA MS Cosette Pelletier

"…She's still in the forward head, with explosive diarrhea and cramps," Mission Specialist Edgar Hodges told the others. Hodges was one of two medically-trained specialists on board, and he was not a little concerned. He had only just arrived on Station the week before, alongside Cosette Pelletier and Alexi Leonov, for the new increment. "Thank God for seat belts on space toilets. She's nauseated, too, but so far hasn't thrown up. It isn't looking good, guys."

"I agree," Dr. Clare Sheehy, the other medico aboard — as well as the Science Officer — concurred. Sheehy had been on Station longer than anyone else, at that point; she was the first test subject of an experiment to enable humans to safely make long-duration space flights. Consequently, the others gave her considerable respect — except for Commander Popov, who tended to give no one much respect. This was starting her eighth increment, so her tenure aboard was closing rapidly on two years. "It sure sounds like Cosette has picked up an intestinal bug. She's been here barely over a week. Someone must have inadvertently broken her pre-mission quarantine, and she brought it up with her, incubating."

M. Brotherton (ed.), *Science Fiction by Scientists*, Science and Fiction,
DOI 10.1007/978-3-319-41102-6_9

"Zhat is indeed…not good," a thoughtful Commander Kazimir Popov decided, his English heavily accented by his native Russian. He had been formerly a MiG pilot for the Soviet Union, leaving the military with the rank of colonel…a fact he had not forgotten. And it showed in the way he ran the increment. While friendly enough, if he issued instructions, the rest of the crew discovered quickly that he viewed them as orders. And, while the revamped RosCosmos did not officially recognize military ranks, in practical application it was a different story.

"Not good at all," Hodges reiterated. "Not only is she becoming dehydrated fast with that much diarrhea, the rest of us risk catching it from her. If more than a couple of us get sick, we'll be in deep shit — pretty much literally."

"Chyort," Popov muttered. "Cannot command if in head." Sheehy threw him a look of distaste, bordering on revulsion.

"How much danger are the rest of us in?" JAXA mission specialist Riichi Maki wondered. "From contracting it, I mean." Though a Japanese citizen, his mother was American, and he had been raised bilingual; his English had no accent.

"Depends what it is," Sheehy decided, pulling her gaze away from scowling at Popov to focus on Maki. "Some of that shit — er, excuse the pun — is actually surprisingly hard to get. Other strains, blink twice and you're puking."

"Dr. Sheehy, you are science officer and ranking physician aboard," Popov addressed the American with officious formality. Sheehy tried not to roll her eyes. "Vhat do you recommend?"

"I hate to say it, but it's my considered opinion that the wise thing to do is to ship her home right away, where she can be properly treated and before the rest of us have a chance to be infected," Sheehy declared, then rubbed her knee as if it ached. "Edgar, what do you think?"

"I concur, Clare. But she'll need someone to nurse her until she gets on the ground. I know you're still involved in that long-duration spaceflight study, so I'll ride down with her, then catch the next flight up if they'll let me bump somebody."

"Good man, Ed. I'll put in a word on the ground, try to get you that bump."

"Thanks, Clare. But I'll need someone to handle the spacecraft; I expect I'll have my hands full with Cosette, especially if she adds vomiting in with the diarrhea in the meantime. She's pretty nauseated, she just hasn't barfed yet."

"Ai! You have placed zhe barf bags within her reach in zhe toilet, yes?" Flight Engineer Alexi Leonov, who had been quiet to this point, finally spoke.

"Yeah, I did. And they all have paper towels folded in the bottom to absorb liquids and prevent splatters. The last thing we need is contaminated barf floating around the cabin. We'd never get everything decontaminated."

"Mm," Maki hummed to himself. "We've gotten behind on that, what with turning over the various increments."

"We have," Sheehy agreed. "It might be good to get on that with the rest of 'em, pretty soon."

"Yes," Maki said.

"Do you suppose ve have already been contaminated, from qvarantine wiz her?" Leonov queried.

"Probably not," Sheehy said, confident. "Generally the person has to be either already blowing chunks — either direction — or right on the verge of doing so, to be contagious."

"All right. If ve take zhe *Soyuz*, rather zhan zhe *Dragon*, I vill ensure you both reach zhe ground safely," Leonov offered. "I am, as yet, still feeling unsure of my skills in zhe *Dragon*."

"I'd rather take the *Dragon*," Hodges remarked. "It has more room, it's a bit faster, and it has a rudimentary medical capability. I'll need all of that; I'm going to try to set up one of the new IV pumps on Cosette, to try to replace some of the fluids she's losing. Otherwise we could have a problem when she hits the re-entry g-forces."

"Uhm. You have a point." Leonov pulled a face. "Yes, zhen ve shall take zhe *Dragon*."

"Very vell zhen. Alexi, check out *Dragon* at vonce. Hodges, prepare your patient," Popov ordered. "I vill call Mission Control in — eh, vherever ve are flying over — and notify zhem of emergency medical evacuation."

Maki glanced out the viewport near him. "Uh, looks like it'll be Houston." He pointed. "Western Hemisphere."

"Houston, zhen." Popov nodded. "Now go."

"Yes sir." Leonov saluted his superior.

"On it," Hodges replied.

"Ed, there's a fresh batch of incontinence briefs in the EVA stowage," Sheehy said, rubbing her knee again. "I'd recommend helping her get into one of those to, uh, 'catch things.' It might be difficult to do, though, if she's still having the explosive diarrhea; if you need help, yell and I'll come."

"Right," Hodges said, pushing off and heading for the *Tranquility* module to see to his patient. "I'll grab some on my way," he called over his shoulder.

"Thank God for the head in the *Zvezda* module," Sheehy muttered in his wake. "The rest of us can still go without dealing with shit everywhere."

"But, um, the…smell…is still a little rough," Maki noted, wrinkling his nose.

"No argument there," Sheehy concurred. Popov snorted in derision as he headed for the *Harmony* module to call Houston.

The others went back to work.

***

Maki returned to the crystal growth experiment on which he had been working when the emergency crew meeting had been called to discuss Pelletier's condition. His hands in a negative-pressure glovebox, he carefully drew off some of the growth solution in a pipette, injecting it into the sample container of the auto-titration unit. Then he reached for a container marked, "Phenolphthalein, Aqueous." The substance was a marker, essential to determine the pH of the solution periodically throughout the experiment, in order to ascertain the precise chemistry of the crystal system as it developed. And the titration unit was equipped with photometric sensors designed to detect the characteristic marker's color, from pale pink all the way to deep purplish fuchsia.

But to his surprise, the container of reagent was nearly empty.

Chikushō, he thought. *What happened to it? That was a new bottle!* He quickly checked the experiment log; there was no record of anyone spilling anything. *A manuke on the other shift made a mess and did not record it, so he —* or she, he considered, realizing that Cosette could have done it as she grew ill *— would not get into trouble for the waste. Or maybe she just didn't have time to record it after she got it cleaned up, if she ended by having to rush to the toilet.* He sighed. *I will have to notify the Payload Operations Center in Huntsville that we will need extra sent up in the next supply mission. And when I run out, we will have to put the experiment on hold until the supply ship arrives. Damn. They will NOT be happy.*

He went to the chemical stowage unit and fetched a fresh, sealed container, easing it through the airlock of the glovebox, then continued work.

***

:::PLAYBACK COMMENCING:::
Houston TX, JSC Astronaut Office, Chief Astronaut's Office
Pre-Increment 50
T minus 45 days, 17 hours, 56 minutes
Subject: MALTSI

"Congratulations, Dr. Sheehy," Matthew Rodriguez, Chief Astronaut, told the physician. "The call has been made, and you'll be the first member of the Mars Long-Term Spaceflight Initiative!"

"Thank you, Matt," Clare Sheehy smiled at Rodriguez, pleased. "I'm honored to be selected as the first candidate for MALTSI. Does it begin immediately?"

"It does. We already have your baselines, and we'll be monitoring you throughout the rest of your prep and into launch. You'll be expected to stay up for about two years, possibly a bit longer, while we evaluate the new hormone protocol. It looks really promising. I have the feeling it will not just slow bone loss. I'm betting that, with a bit of adjusting of the dosages, it'll eventually prevent it altogether."

"I've been following the studies and trials, and professionally I agree," Sheehy replied.

"Excellent! Have at it, Clare! You're going to make us proud," Rodriguez declared.

***

:::PLAYBACK COMMENCING:::
Star City, Russia, Moscow Oblast
Increment 50
T plus 57 sec
Subject: Launch of First MALTSI Test Subject

"…Telemetry shows Dr. Sheehy's vitals are nominal for ascent phase," the flight surgeon determined, carefully watching the numbers that flashed across his computer terminal's screen.

"And she was in peak condition prior to launch," the MALTSI mission scientist pointed out over the comm headset from her location in Huntsville, Alabama at the Marshall Space Flight Center.

"True. Once docking and transfer are complete, we are go to begin MALTSI protocols."

***

:::LIVE DOWNLINK COMMENCING:::
International Space Station
Increment 58
Mission Elapsed Time 8 days, 11 hours, 04 minutes
Subject: Decontamination

After the trio had departed the Station on the medevac mission, decreasing the crew complement by half, a thought had apparently occurred to the commander, and he located Maki.

"Riichi, vill you have zhe good place in zhe experiment, any time soon, to pause for a little vhile?"

"Well, yes, but it will be a few minutes yet. What do you need?"

"It has occurred to me zhat ve most likely vant to get zhe trash stowed in *Progress* and sealed avay as soon as possible. And zhey should receive hazardous biological vaste labeling, in contrast to usual labeling, so ve do not shift around anymore zhan needed later. But I cannot handle it, as I am still coordinating viz Mission Control in Houston; zhey have brought Pelletier's physician to ESA's Columbus Centre Control. Huntsville has also been notified, so zhe replan to experiment timeline, to accommodate zhe reduced crew complement, is in vork."

"I had rather not, Kazimir, if it can be avoided. I am behind schedule on this experiment as it is, even if they do replan the activity. We're doing good to get any experiments done at all, with half the crew gone, and I really want to at least finish this run. Plus, I just discovered that Cosette spilled reagent into the glovebox during the initial onset of her intestinal virus, and we will be low on supplies for the experiment. And I shall have to use more experiment supplies to clean the galley of any contamination, unless I am authorized to use the medical supplies for gloves and the like. Cannot Clare do trash duty?"

"Dr. Sheehy INFORMED me zhat she is sanitizing zhe toilet in *Tranquility* module," Popov explained, pulling an annoyed face.

"Eh. I take it she was her usual, ah, blunt, self."

"She vas, indeed. Blunt is understatement. I zhink she does not like me."

"I'm not sure how much she likes any of us," Maki admitted, "but then, she's been up here so long, she's probably sick of it in general. I wonder… you know, when we first came aboard, last increment, remember how she was always up in the cupola, looking out? Then suddenly she just…stopped. And now she avoids the viewing ports like the plague. She seems…ill-tempered… most of the time now, too. Could it be homesickness?"

"Hm. Zhat is…qvite possible. I vill discuss it vit zhe surgeon at Star City vhen I have next medical conference. He may advise. I do not like insubordinate crew." Popov gave Maki a stern glance. "But back to business. She is cleaning zhe head. It vill no doubt take some time, and I felt it best not to vait any longer for infection to fester in zhe general vaste. It must be removed at vonce."

"Uh. Good point. Okay, give me five and I shall see to it."

"Excellent. Use vhatever supplies you feel are needed to do zhe job properly, and protect yourself from contamination. If our cantankerous yet excellent Science Officer objects, fear not — your commander shall handle zhe matter."

***

Maki had completed crew operations on the crystal growth experiment, initiating automated ops. Then he had fetched latex gloves, safety glasses, and a face mask from the medical supplies, donned them, and begun the unpleasant task of gathering the galley waste, choosing to do that first, then check the sleeping berths for trash which might not be there, but which would certainly be more contaminated than the galley waste. *In addition,* he considered, *I probably should look into placing her uniforms into a hazardous waste bag, as well. We do not need to have it migrating around among our own clothing.*

Now he studied the trash bin in the galley, considering whether or not to load it into a wet trash container. But that would require venting any outgassing into the waste management system, which could potentially re-contaminate what Sheehy was working so hard to decontaminate now. *No,* he decided, *I'll double-bag it and seal it, put it in a hazardous waste bin, and that into the* Progress, *and be done.*

So he pulled the bag out of the trash receptacle and gingerly shook it down, knowing he'd have to clean the whole galley area very thoroughly with alcohol wipes when he was done.

Maki was about to close and seal the bag when something caught his eye: a bright fuchsia color in the waste.

"Nande kuso?!" he exclaimed in surprise, then poked around in the garbage with gloved hands. Soon he brought out a container and a water syringe — the container had held pasta Alfredo, the sauce residue of which was now a bright pink, the color that had caught his eye. The syringe had been used to rehydrate the Alfredo sauce, but it, too, had a slight pinkish tinge to the remnants of the fluid inside. And the color was unmistakable, especially since he had been working with it so much for the last month.

Phenolphthalein, he thought. *Cosette didn't spill it — she ATE it. She mixed her sauce with it instead of water. No wonder she was sick! Besides, that should have tasted of ethanol to the skies! Ieuch! But,* he remembered, *she has only just arrived and is probably still suffering from SAS, so her taste is likely off. Plus, she is from France, and was trying out the new wine containers from the French vintners, so that might have hidden the taste. But why in heaven's name would she do that in the first place?!* Then a possible answer came to him.

Pelletier, several of them had noticed, had not dealt as well with life aboard the Station as most astronauts and cosmonauts did. She had grown up living on a vineyard, was used to open spaces, and the confined volume of a space station — especially when coupled with the initial disorientation of micro-gravity — had unexpectedly produced some level of claustrophobia in her, especially during her first sleep period. She had gotten essentially NO sleep for the first night or two. Maki knew that for a fact, because she was still on his shift for the first couple of days, and her berth was right above his; he had heard her panicked breathing most of the first night. Halfway through the sleep period, he'd asked her if everything was all right, and she admitted that the sleeping berth seemed far too small, and it made her uncomfortable. They had both tacitly known that she was grossly understating her reactions.

Maybe, he considered, *she couldn't take it, and thought this was a way to get home fast, without it looking too badly on her curriculum vitae.* He shook his head. *That's not good, on several levels. I wonder if I should tell Popov.*

Then something else occurred to him, of immense importance. *The Soyuz. If I'm wrong, and Cosette really was sick, and this is just some food dye gone bad — in which case she possibly suffers from food poisoning, just as bad as an intestinal virus — then if the lot of us get sick, it would be good to make sure that the* Soyuz *is in a good way to get us all safely home with as little manual intervention as possible...*

Quickly Maki gathered up the bag of trash, setting it to one side to show Popov later, and headed for the *Rassvet* docking module, peeling off his gloves, mask, and goggles as he went.

***

:::PLAYBACK COMMENCING:::
International Space Station
Increment 56
MALTSI Mission Elapsed Time 1 year, 6 months, 24 days, 9 hours, 27 minutes
Subject: MALTSI Test Subject Fitness
Sheehy unstrapped from the platform and pushed out of the densitometer scanner, floating over to the monitor to look at the specialized x-rays of specific parts of her skeleton — lumbar and thoracic spine, right pelvic girdle, and left forearm. She tapped a couple of buttons on the computer console; this sent the imagery to the MALTSI science team on the ground, in the Science Operations Area of the Huntsville Operations Support Center. They would view the images at the same time she did.

Abruptly the images popped up onscreen, and she gasped.

"No, no, no," she whispered, staring at the x-rays in horror. "No, it can't be. I doubled the dosage!"

"MALTSI Huntsville to ISS on Air-to-Ground 2. Sheehy, are you seeing this?"

"Sheehy to MALTSI Huntsville! Affirm! What the—" she caught herself before she cursed on Air-to-Ground — a huge protocol violation — and amended, "what the heck is going on?! I've followed the protocol to the letter! I even doubled the hormone dosage like you recommended!"

"Yes, I know. We had your electronic logs," Chris Adams, the MALTSI Chief Scientist, replied. "Evidently there is something radically wrong with the protocols. I—"

"Break-break. Flight on Air-to-Ground 2. POD, INCO, GC, take this conversation to the private loop for a medical conference. Surgeon, stand by for PMC."

"Flight, GC copies."

"Flight, INCO copies. Stand by one…"

"Flight, Surgeon, standing by."

"Flight, POD copies. DMC, attend Flight loop."

"DMC copies."

Sheehy wanted to fling something while she waited for the go-ahead.

"Surgeon, ISS, MALTSI, this is INCO. Comm switched to AG-3; recording off, loop locked. You are go for private medical conference."

"Thank you," Adams' voice answered. "Clare, we are honestly unsure of what has happened. The increased dosage appears to have ACCELERATED bone loss, rather than slowing it, as we expected."

"I NOTICED THAT!" Sheehy practically screamed into the mic. "Did you happen to notice the microfractures in my forearm?"

"We did," came the answer. "By any chance is that the arm you use for pushing off and deflecting, as you move about the cabin?"

"It is," she whispered in dismay, comprehending. "Stress fractures. My bones are brittle." She opened her mouth to say more, but what came out was more nearly a whimper, so she stopped, struggling against tears of fear.

"Break-break, Flight Surgeon on AG-3 for Sheehy."

"Go, Surgeon."

"Dr. Sheehy, this is Pete Caldwell. We have a serious situation here. I don't think any of us expected a bass-ackwards response to the protocols, but that seems to be what has happened. I've been following the experiment very closely, and I am exercising my authority as chief flight surgeon for the increment to call a halt to it, effective immediately. It is my considered opinion that continuing it any further puts your life at risk…if it hasn't done so already."

"Surgeon, Sheehy. What do you mean, 'If it hasn't done so already'?"

"Clare," said Cardwell, and Sheehy knew by the soft tone of his voice and the personal mode of address that what was coming wasn't good. She was right. "Clare, right now, I don't think your skeletal structure could handle your own weight on the ground. I'm really not sure you could handle the g-forces involved in a re-entry. It could very well kill you."

"Damn," Sheehy breathed, careful not to key the mic on her comm headset. A cold hand seemed to reach into her gut and grab a handful of entrails; she shivered. Several tears spilled over and bubbled around her eyes, leaving her unable to see. She dragged her sleeve across her face, absorbing the liquid and clearing her vision. Finally she keyed the mic, since she heard nothing but silence on the other end. "What should I do, then?"

"You stay put and keep your chin up," Cardwell answered. "We'll figure something out. And Clare, Dr. Adams…this stays between us. Not even the other crew, Clare. You know how paranoid some of the Russians can be about medical matters. It's a cultural thing, but it would be bad for morale. Besides, it's your personal medical information, which is nobody's business but yours and the medical personnel treating you. Do you understand?"

"Affirm," Sheehy mumbled into the mic, deeply troubled.

"…Affirmative," Adams murmured. "Dr. Sheehy, our deepest apologies… I…we…we have no idea how this backfired so badly."

"Surgeon, Sheehy. What protocols do you want me to follow in the meanwhile?"

"Increase calcium and phosphorus intake, boost vitamin D, try to exercise if you can without causing stress fractures. I'll work on it from this end. Once we can figure out how to boost your bone density a sufficient amount, we'll start you on THAT protocol, and get you home when it's built you up enough."

"So I'm stuck here in orbit until then," she declared, disbelieving.

"…Yes. For now. We'll come up with something, Clare, I swear."

***

:::LIVE DOWNLINK COMMENCING:::
International Space Station
Increment 58
Mission Elapsed Time 8 days, 12 hours, 02 minutes
Subject: Galley Waste
Popov and Sheehy arrived in the galley at the same time. Sheehy was looking for something to drink after thoroughly cleaning the toilet facility in the Tranquility module, and went straight for the water. Popov was looking for Maki. Finding the trash stowage only partly completed — barely begun, in fact — he frowned in annoyance.

"Govno," he grumbled. "Vhere is Maki?"

"Dunno," a curt, out of sorts Sheehy replied, sucking in a deep pull of the water. "I been up to my elbows in 'govno,' so I haven't seen him. You check the materials lab?"

"No, I — vhat is zhis?" Popov gingerly pulled open the top of the stowage bag and peered in. "It is…pink. It is food, but it is…pink."

"What?" Sheehy wondered, pushing off and floating over. "Pink? You or Alexi been eating borscht?" She smirked.

"No," Popov replied, brusque, "and is not right shade pink, anyvay. Is…is bright pink. Look."

Sheehy peered over Popov's shoulder.

"Damn. It really IS pink, isn't it…?"

"It is indeed…" Popov murmured, staring into the bin. "Very pink."

Sheehy eyed Popov with a distrustful gaze.

***

:::PLAYBACK COMMENCING:::

International Space Station

Increment 57

MALTSI Mission Elapsed Time 1 year, 8 months, 15 days, 4 hours, 12 minutes

Subject: Public Affairs Office/ISS Video Opportunity

The ISS Increment 57 crew — Commander Popov, Flight Engineer Peter Murphy, and Mission Specialist Riichi Maki, along with the Increment 56 crew of Science Officer Sheehy and Mission Specialists Alan Cocoran and Maksim Vasileyev — had gathered during shift handover for an awards show ceremony, which was to be filmed and downlinked to Los Angeles.

All went well until the very end. Since the award was for a science fiction category, the awards people wanted something special from the astronauts. So Popov had decided that they should all turn backflips in the microgravity environment. Sheehy had argued the matter, suspecting that so many in such a tight volume would produce problems she didn't want to deal with.

"Kazimir, it isn't safe, not with all of us. We have racks on either side, the stowage below, cabling…it's an accident waiting to happen."

"It vill be fine," the commander pressed, stubborn.

"What if someone kicks — er, one of the control panels? It could reset entire experiments! The ground will be twelve kinds of pissed. Or somebody could get their feet tangled in the cables!" Sheehy was careful to avoid mentioning the idea of someone hitting another person — that came perilously close to her secret, and she had express orders not to discuss THAT.

"Ve vill simply tie down zhe cables before zhe downlink," Popov shrugged. "Ve are cosmonauts and astronauts, Doctor. Ve know vhat ve are doing."

"But—"

"Zhat is enough, Doctor," Popov cut her off, frowning. "Zhe event will go forward as ve haff planned it."

"Because?" She put her hands on her hips in defiance.

"Because I say so. And I am zhe increment commander." And that was the end of the argument. Mission Control Moscow had already shown a tendency to support a strong chain-of-command structure, and Houston was unlikely to override it. The synchronized backflips were a go.

Unfortunately, as they twisted in midair during the broadcast, exactly what she had feared came to pass: Cocoran kicked Sheehy in the head.

***

In the alcove of the *Columbus* module reserved for medical treatment, a dizzy, wobbly Sheehy stared at the computer screen. On it was an x-ray that depicted a hairline fracture of the left parietal bone of an adult human female skull. She ran a trembling finger over the line denoting the fracture.

Then she gingerly palpated the left side of her head, where Cocoran had kicked her, and winced.

***

:::LIVE DOWNLINK COMMENCING:::
International Space Station
Increment 58
Mission Elapsed Time 8 days, 11 hours, 56 minutes
Subject: *Soyuz* Space-worthiness
Inside the *Soyuz*, Maki checked the settings as carefully as he could. He'd had the standard Station emergency egress training, but had chosen to go beyond that, and study what he could coax out of his cosmonaut colleagues, learning what he could about actually flying the *Soyuz*. He had also been working on the same matter with the *Dragon* craft, but was not as far along.

So it was the matter of a few glances to realize that the settings were not what they ought to have been for a docked craft in standby mode. *In fact,* he thought, *this looks like prep for an imminent deorbit burn. But the docking clamps are locked in place. I didn't even think you could SET a burn with the clamps locked. I thought you had to be free-floating.*

He opened an access panel nearby to see if there was a problem with the engine connections and was shocked to find several wires had been cleanly snipped. *Well, that explains it,* he thought, grim. *That's...that was...deliberate.* He pushed back, being careful not to fling himself across the cramped cabin into the far bulkhead, and considered. *That means...someone on board, one of my crewmates, is trying to sabotage...*

And suddenly the phenolphthalein stain in Cosette's meal made sense.

But why? And who? And why Cosette? There are only two other crew members on the Station, so I have a 50/50 chance of being right, whichever one I pick. He pushed down into one of the seats and pondered. *With the increasing usage of the* Dragon *capsule, the Russians no longer have a monopoly on travel to and from the Station. That could mean a political reason to do something. And Leonov wanted to take the* Soyuz, *not the* Dragon. He ran a distracted hand through his hair, standing it on end, where it waved gently in the slight air currents. *I need more data to figure this out. But in the meantime, I need to do something here.*

He pushed out of the seat and floated slowly over to the controls. Reaching inside the access panel, he matched wires, then opened one of the cargo pockets on his uniform and fished out a multi-tool. He stripped insulation off the cut ends, then spliced the wires back together. Closing the panel, he flipped the switches back into standby mode.

Then he peeped outside the hatch, into the docking module, to see if anyone was watching. No one was, so he exited the *Soyuz* with a hard push, heading for the nearest communications console to call a mayday on the air-to-ground loop as fast as he could.

He didn't see the wire cutters that drifted out from under the command console behind him, nor the countdown timer slung underneath it.

***

:::PLAYBACK COMMENCING:::
International Space Station
Increment 57
MALTSI Mission Elapsed Time 1 year, 10 months, 2 days, 6 hours, 38 minutes
Subject: MALTSI Test Subject
Sheehy crouched at her desk in the medical alcove of the *Columbus* module, her feet hooked securely into the anchor loops, so she wouldn't drift away. She studied the latest densitometer x-ray of her femur, and scowled at the hairline fracture near the knee. Unconsciously she rubbed that joint, which had begun to ache a month earlier, when she had first discovered the break. It did not appear to be healing properly. She shook her head, chewing her lower lip in thought.

At least the skull fracture has healed. She pulled up the latest image of her skull, which showed only a faint shadow in the parietal bone where the crack had been: the new bone was less dense than the old. *But,* she thought, *that's to be expected; it's the way the bone grows back after extended weightless conditions. It isn't going to get any better until they come up with a proper treatment that WORKS.*

Just then, her headset beeped, and CapCom announced, "ISS, Houston for Sheehy." She keyed her mic.

"Houston, this is Sheehy. Go."

"Please switch to AG-3 for a private medical conference."

"Copy; switching to AG-3." She hit a couple of buttons on the comm console sitting beside her laptop. "Houston Surgeon, this is ISS Sheehy for requested PMC."

"ISS Sheehy, this is Surgeon. How are you doing, Clare?" It was Dr. Cardwell's voice.

"About as well as can be expected. I guess you see the stress fracture just above my knee."

"We do. And the healed skull fracture. What happened there, and why didn't we know about this sooner?"

"Uhm," Sheehy began, "I…didn't think it was necessary. I didn't show any signs of concussion, though I told the others I did, so I could bandage it and pad everything." She paused, then added with some bitterness, "It isn't like you could do anything if you DID know."

"So…you've seen no after-effects? No long-term or permanent damage?"

"Negative. Hurt like hell at the time, and I had a headache for a week, but it healed up fine."

There was silence on the private air-to-ground for several seconds before Cardwell continued.

"…How did it happen, Clare?"

"We had to do that public affairs crap for the awards show on TV, few weeks back," she snarled. "They wanted something suitably 'spacy,' and Popov got the bright idea we should all turn backflips in the cabin, all at once. 'Synchro- uh, synchronized flipping,' the damn TV moron called it. I TOLD Popov there wasn't enough room, that something would happen, but he wouldn't listen — he pulled rank on me, the offi- uh, the stuck-up moron — and sure enough, it did. Cocoran kicked me in the head. Blame them. I sure as hell do."

"I'm sure it was an accident, Clare," Cardwell soothed, apparently ignoring the cursing, not that she cared; it was, after all, a private communique. "Cocoran didn't mean to hit you. And neither he nor Popov could have known your…condition. Unless you told them."

"It was a stupid idea!" Sheehy shot back, venom in her voice. "I told that moron Popov not to do it!"

"But not why."

"I shouldn't have needed to! I told him it was too close, there were controls, wires, cables, all kinds of things that could be messed up by a stray foot, knee, elbow, whatever! But no! 'Zhe show must go on, because I say so!'" she mimicked the Russian's accent. "And I have no idea how Cocoran managed to make it through to the astronaut corps without washing out. He is the most uncoordinated, insensitive, obliv- uhm, blind-to-his-surroundings EXCUSE for an astronaut I have ever seen!"

"Calm down, Clare." Cardwell attempted to appease the irate astronaut. "You can't blame them for what they didn't know about. And I expressly told you not to tell any of 'em."

"True." Sheehy took a deep breath and let it out. It didn't really seem to help that much, but at least it enabled her to sound calm. "So do you have anything for me? Something that'll get me home soon?"

There was another pause, and she tried not to grind her teeth, well aware that it could cause damage, not just to her teeth at this point, but to her jaws. *And that WOULD be a mess,* she thought. *Break my jaw, then I can't even manage to cram in what nourishment I'm getting, and the situation deteriorates further. And I can't get home to medical treatment, so I'd have to wire my own jaw shut. Fun, fun.*

"The news is bad, Clare," Cardwell finally admitted. "When the MALTSI interim report came out, the National Institutes of Health got fourteen kinds of livid, and cut off MALTSI's funding at the ankles. The program is gone, and without it, we have nobody doing the research we need to get you home. I'm working on it hard; I've pulled in the Chief Astronaut, the Chief Medical Officer, even the Deputy Administrator, who is considering approaching the Chief Administrator in confidence. We're trying to scrape up the monies to replace the NIH funding for MALTSI so the research can continue."

Sheehy fairly gaped at the comm panel in horrified shock.

"…And if you can't?" she eventually choked out.

"Don't even think it, Clare. We will. We have to. We'll think of something."

"I'm going to die up here, aren't I?"

"No, Clare. Not if I can help it." Cardwell could be heard swallowing, then he said, "Keep your chin up, kiddo, and we'll talk as soon as I know anything more." And he broke the comm.

Sheehy sat, staring blankly at the console, rubbing her knee absently.

Gradually, first her eyebrows, then her entire face, drew together into a fierce, angry scowl.

***

:::PLAYBACK COMMENCING:::

Houston TX, JSC MCC Medical Back Room

Increment 57

MALTSI Mission Elapsed Time 1 year, 10 months, 2 days, 7 hours, 12 minutes

Subject: MALTSI Test Subject

Flight Director Gayle Murphy and defunct project MALTSI's chief scientist Dr. Chris Adams sat across from Peter Cardwell. Their faces were solemn.

"That…was not good," Adams remarked, forehead creased in worry.

"No, it was not," Cardwell agreed, equally concerned. "And it was not at all like the Clare Sheehy I know, either."

"Me either," Adams confirmed.

"Meaning?" Murphy wondered.

"Meaning there's something going on in her head, and it doesn't bode well," Cardwell explained. "What's worse, she doesn't see the damage."

"What damage?" Murphy asked. Cardwell looked past her at the MALTSI alumnus. Adams nodded at Cardwell, and reached for his laptop, waking it and pulling up some images.

"Look here, Flight," he murmured. "You heard Pete talking with her about the skull fracture."

"I did, and I found it most alarming."

"As well you should," Cardwell said. "See these two x-rays? This one," he tapped Adams' laptop screen, "is from Clare's pre-flight checkout. This other one is the image she just downlinked, though I'm sure she forgot it was in the batch dump."

"Oh, dear God," Murphy whispered. "She's got a…a dent, in, in her head."

The two scientists nodded.

"I saw that right off," Cardwell noted. "That's a significantly depressed area in the parietal region of the skull, resulting from the improper healing of the skull fracture, and I cannot for the life of me see how, given that level of damage to the skull, there hasn't also been damage to the brain beneath it. The parietal lobe, specifically."

"Damnation," Murphy said, staring at the screen. "And so you think that her, her…"

"Her personality changes," Adams supplied.

"Yes, her personality changes—"

"—Are almost certainly due to the brain damage," Cardwell finished. "Yet she can't see it, or recognize that she had to have had symptoms of concussion. More, that region of the brain is involved in proprioception, the ability of the brain to coordinate spatial sense and navigation, meaning she's much more likely to slam into things in the close confines of the cabin."

"Which in turn causes increasing damage to her skeletal structure," Adams added.

"And it's also involved in language processing. Did you notice how she stumbled over some of her words, and got certain others — like 'moron' — stuck in her head?" Cardwell asked. He paused, tapping his fingers to his lips as he considered the situation. "We have a seriously compromised crew member."

"Then let's bring her down, get her medical help."

"We can't," Cardwell replied, maintaining an impassive façade. "The whole reason for this problem in the first place is because the bone density protocols backfired somehow. Right now Clare Sheehy couldn't sustain her own body weight on Earth — her entire skeleton would collapse and it would kill her. Let alone the g-forces encountered during re-entry."

Murphy stared at them, flabbergasted.

"What the bloody hell are you doing about it, then?"

"We're working on it, Gayle," Cardwell told her, trying to remain calm, running his hand through his hair as a distraction. "That's why Chris is here. I brought him in from MALTSI's home office. I had to play fast and loose with some rules, but I hired him to consult, at least temporarily, with my private practice. That at least covers his travel expenses."

"Meanwhile, the MALTSI team is working — without pay — to figure out what went wrong, and why, and try to determine what to do about it," Adams added. "We think we have it narrowed down to some nepotism at the university, between a graduate student and his advisor that shouldn't have been his advisor, and some totally falsified data." He shook his head in disgust. "Damn sloppy excuse for science. I am NOT pleased, and I and my immediate colleagues are working to get the advisor fired, and the student — his nephew — expelled. YES, it was that bad. It was unconscionable. And I shall never forgive myself for being part of this whole…FUBAR. Hell," he added, bitter as quinine, "I ran the thing. The buck stops here."

"Ease up, Chris," Cardwell murmured. "Big science project like that, you can't do all the research by yourself. At some point, you have to trust your colleagues to do their jobs right."

"Yeah, well, they didn't," Adams said, blunt and harsh. "It's our fault, and I damn well know it."

Cardwell had no answer for that. The three sat quietly for several minutes, just staring at the top of the conference table, worried.

"At any rate, we think we know where the mistake came in, and how it got improperly validated," Adams admitted. "And we have some ideas about how to reinterpret the hormonal activity in bone catabolism and anabolism. But…"

He broke off, eyes going distant, and the other two waited silently for him to finish. When Adams resumed, it wasn't what either flight controller wanted to hear.

"…But we don't know if we can help Dr. Sheehy. She…may be too far gone."

"SHIT," Gayle Murphy said with feeling.

***

:::PLAYBACK COMMENCING:::
International Space Station
Increment 57
MALTSI Mission Elapsed Time 1 year, 10 months, 3 days, 7 hours, 44 minutes
Subject: MALTSI Test Subject

"…So there's nothing you can do? I'll really die up here?" Sheehy whispered into the mic, on her second PMC in as many days.

"I didn't say that, Clare," Caldwell countered. "I said there's nothing we can do right now. We're not giving up, and you mustn't, either. Just hang on. Give us some time."

"Dammit, Pete! Do you know how long I've been up here already?! Up here with these idiots and morons?"

"Clare! Calm down! We're doing everything we can! Now, I'm going to send up some new instructions, and I want you to follow them to the letter…"

***

:::LIVE DOWNLINK COMMENCING:::
International Space Station
Increment 58
Mission Elapsed Time 8 days, 12 hours, 14 minutes
MALTSI Mission Elapsed Time 1 year, 11 months, 18 days, 12 hours, 16 minutes
Subject: Sabotage

Maki left the *Rassvet* docking module, headed for the nearest communications console, but grabbed the closest handhold to stop as soon as he entered the *Zarya* module.

In front of him hovered Clare Sheehy, an evil leer on her face, and a wild light in her eyes. Her hair was free of its customary ponytail, and floated about her head in unrestrained disorder. Somehow its random, rippling patterns, as it appeared to stand on end, only served to heighten the deranged expression on her face. ·

Her left hand dangled awkwardly from a joint in the middle of her forearm that shouldn't have been there. In her right hand she held a large mallet which Maki recognized as being from the Station's EVA tool kit. To his horror, it was smeared with what looked like blood and some sort of yellowish-gray tissue. Red globules wafted in her wake, leaving a gruesome trail behind her; peering past her toward the *Zvezda* module, he saw the limp body of Commander Popov, glassy eyes staring, unblinking. As the commander's body drifted in the connecting passage, it pivoted slowly, and Maki saw the dreadful wound on the back of his head, where his skull had been smashed in. And Maki knew what was on the mallet Sheehy wielded.

He swallowed hard, forcing bile back into his stomach from where it had crept into his throat, and returned his attention to Sheehy. Her leer had deepened.

"I told him we shouldn't do it," she said. "That it was dangerous. But he insisted. So I paid him back, tit for tat. Cocoran — he's the one who really did it, you know. He got off the Station before I could return the favor for it. But you? You're still here. It was Popov, and Cocoran…and you."

"Clare, I have no idea what you're talking about," Maki protested, edging backward toward the *Soyuz*, wondering if he could take refuge behind the closed hatch, then depart the Station alone. "Please, calm down and put away the mallet. Let's just talk over what's going on here. I'm sure we can find a way to fix things."

"Oh, we're going to fix things," Sheehy declared. "They told me I couldn't go home. But they're wrong! I'm going home, as of today!"

"So are you the one who reset the controls onboard the *Soyuz*? I did a quick check just now, and they weren't in standby mode anymore."

"You found that, did you?" She crept closer. He eased back.

"Yes, but it was dangerous, so I reset everything to standby mode."

"Think so?" Sheehy smirked. Maki hid his cringe at the demented expression.

"Yes. I even spliced back the wires inside the port access panel. It's safe, Clare. We're both safe. Let's keep it that way. I'll call down, we'll get you help, we'll get you back home and everything will be okay."

The laugh that emerged from her throat had more than a hint of hysteria in it.

"Get me back home! They'll help! Those damn morons! They can't even help themselves out of the rain! But I don't need their help any more. I'm making my own way home, Riichi. Along about…now."

There was a sudden loud thump that rattled the entire Station, followed by a kind of muted roar that went on and on. Then, to Maki's surprise, he and Sheehy began slowly shifting toward the starboard side of the module.

"Chikushō!" he exclaimed. "That was a *Soyuz* deorbit burn!"

"That's right," a grinning Sheehy said, moving closer. "I told you, I'm going home. I'm going home if I have to take the entire damn Station with me!"

And she kicked hard off the bulkhead, hammer raised. Maki heard a sickening crunch, saw her lower legs seem to collapse as she screamed in agony, still swinging the hammer with all her might.

***

:::LIVE DOWNLINK COMMENCING:::
Houston, Johnson Space Center, MCC Front Room
ISS Final Increment 58
Mission Elapsed Time 9 days, 2 hours, 8 minutes

"…Dear God," Flight Director Gayle Murphy whispered, watching the images on the big screen from the local television station depict the fiery re-entry breakup of the International Space Station over the waters of the Pacific, filmed from several different islands scattered across that ocean. "What has she done?"

"She?" Flight Surgeon Dr. Peter Cardwell replied, standing beside her, sick at heart. "She was dying by inches, Gayle. All for the sake of the dream."

"But…the dream is worth it…isn't it?"

"It is…but not like this, Flight. Never like this…"

"Flight, PAO, your loop. The news media is waiting, ma'am. Along with several Congressmen."

Murphy groaned, pulling off her headset. Cardwell sighed, then turned, taking her elbow.

"C'mon, Gayle. Time to face the music and dance."

:::END DOWNLINK:::

Afterword

I enjoy blending hard SF with mystery, and with my background as a payload flight controller for seven Shuttle missions plus several ISS increments, it seemed a logical notion to set a mystery aboard a near-future Space Station. I had advice from friends and former colleagues Dr. James K. Woosley and Larry Bauer.

I would like to specify I created a communications loop for this story. Normally private conferences (medical etc.), occur on Air-to-Ground-1, locked down outside intended conversants. But since that's the main comm between Houston and the "guys upstairs," I envisioned AG-3 expressly for any private comm from Earth to spacecraft.

For those interested, I've listed some references on subsequent pages for more information, starting with Wikipedia and moving to more sophisticated sources. But to summarize the core issues at the heart of this story…

Phenolphthalein has been in use for over a century as a laxative, though concern over long-term carcinogenicity has caused it to be removed from store shelves in recent years. However, it is also commonly used in chemical titration as a colored marker to denote pH in the solution being tested. I have myself used it for this purpose, and a skilled chemist can determine pH of a solution quickly and accurately with it, without need of additional or more sophisticated instrumentation. But titration equipment is gravity-based, so a space-based lab would need automated equipment. Arguably there are other means of instrumenting a solution to determine pH, but as aforementioned, a skilled chemist is just as fast, and in some cases, more accurate.

It's a concern that long-duration space flight may cause mental issues, even up to and including psychotic breaks, due to the relative isolation and very restricted environment — it is not, after all, as though one can go for a relaxing stroll after dinner! If those factors are combined with brain damage due to impact/skull fracture, serious psychological problems might ensue.

Also microgravity effects upon skeletal systems are known. It seems there's a glitch that develops between systems breaking down old bone and those building new bone, such that in microgravity, the new bone systems significantly reduce activity, if not outrightly shut down, while old bone systems maintain, or even possibly increase, activity. More, it produces permanent damage: while the bones remineralize upon return to a gravity environment, they remain less dense than before the spaceflight. It is perfectly reasonable, therefore, to postulate a point of no return — where bone loss has become so great, the skeletal system cannot survive return to a gravitational environment.

Since bone breakdown/deposition systems are hormone-regulated, a reasonable means of attempting to circumvent bone loss might be artificially manipulating those hormones. However, we're still working to understand this synergy, and what works on Earth might not necessarily work on everyone in space. More, if we add in a postulated data scandal such as has been uncovered in a couple of different fields in recent years, there is the potential for harm over and above the normal spaceborne bone loss.

I'd like to point out that this story is an exploration of possibilities, not an indictment of any system, and certainly not intended to be anti-NASA; I am very pro-NASA. If it is an indictment of anything, it is the tendency in some branches of the scientific community to falsify data to get ahead at the expense of real science.

Ultimately it must be remembered that space exploration is an inherently dangerous business, given the inimical environments surrounding the explorers. However, it is not only desirable, it is NECESSARY. Without doubt there are large rocks out there, quite capable of destroying civilization as we know it. Even lesser rocks have the capability for destroying entire cities; the Chelyabinsk bolide over Russia in 2013 could have wiped out the city had its trajectory been more nearly vertical. And this, without it ever having to strike the ground — the transfer of energy and momentum, the same transfer which shattered windows and knocked down walls over a broad area due to the shallow angle, would have flattened everything beneath it, and possibly produced a shallow crater into the bargain, had it been less tangential in its trajectory.

It therefore behooves us to maintain and expand our spacefaring capabilities, for certainly there are dangers out there…but there are also resources as well, resources that will aid us in expanding our civilization and our capabilities.

Humanity is, in the end, an explorational species by its very nature, and ever have we looked upward to the heavens at night and yearned to be among the stars. Perhaps it is time to acknowledge that we are intended to be "out there," and pursue that goal with more determination, despite potential danger.

References

NASA console positions:

* Flight — Flight Director (JSC)
* Surgeon — Flight Surgeon, chief medical officer on the ground (JSC)
* INCO — Instrumentation and Communications Officer (JSC)
* GC — Ground Controller (JSC)
* PAO — Public Affairs Office/Officer (JSC)
* CapCom — Capsule Communicator (terminology and call sign dates back to early NASA flights) (JSC)
* POD — Payload Operations Director (MSFC)
* DMC — Data Management Coordinator (MSFC)

Crew positions:

* http://www.encyclopedia.com/doc/1G2-3408800279.html
* https://en.wikipedia.org/wiki/Astronaut_ranks_and_positions# International_space_station_positions

Effects of microgravity on skeletal systems:

* https://en.wikipedia.org/wiki/Effect_of_spaceflight_on_the_human_body
* https://en.wikipedia.org/wiki/Spaceflight_osteopenia
* http://science.nasa.gov/science-news/science-at-nasa/2001/ast01oct_1/
* http://www.mc.vanderbilt.edu/gcrc/space/#Bone
* http://www.nasa.gov/mission_pages/station/research/experiments/118. html

Miscellaneous:

* Aerospace Medicine coursework, Section III 3.2 Waste Management, Federal Aviation Administration, Aeromedical Education Division, Civil Aeromedical Institute, 2004, Kira Bacal, MD PhD.
* https://en.wikipedia.org/wiki/International_Space_Station#Mission_controls
* PMC — Private Medical Conference. The Air-to-Ground loop being used is placed into private mode, and the astronaut (who may be ill, or may be on an experimental medical protocol) converses with the Flight Surgeon, with option for more medical personnel to be involved.
* SAS — Space Adaptation Syndrome aka space-sickness. Most have it in some degree on first flights.

Other languages:

* "chikushō" — Oh shit! Oh blast! Oh hell! (Japanese)
* "manuke" — Clueless, loser (literally "out of rhythm" or "missing a beat") (Japanese)
* "Nande kuso" — What the hell? (literally "Why shit?") (Japanese)
* "Govno" — Shit (Russian)
* "Chyort" — Damn, hell (Russian)

One for the Conspiracy Theorists

Jon Richards

I am a scientist at the SETI Institute searching for radio signals hinting at extraterrestrial life. We're not UFO seekers, we're just serious engineers using hard science to detect a radio signal from anywhere that hints we are not alone in our galaxy.

When I am forced to attend a social event or party (engineers hate parties), I am inevitably asked, beer in hand, "So, what do you do for a living?" If I answer honestly, I get one of two responses. If the person has never heard of SETI, they look at me like I am a UFO conspiracy nut and try to inconspicuously slink away. The other response is wide eyes and the excited question, "Find anything yet?" Before answering, I think to myself, "Well, it's complicated; how much time do you have?" Then I smile and say, "Not yet."

The truth is *most* of the time I don't see any of those radio signal hints. Like I said, "It's complicated…"

Is there a civilization on a planet emitting radio signals that we can detect? We won't know if we don't look. I'm the guy looking. I am looking. I'm lucky enough to be the operator of the SETI radio searching for a needle in an infinity of haystacks at the Allen Telescope Array.

The ATA is located in remote northern California among cow pastures and ancient lava fields at 3,200 feet elevation. The SETI Institute owns forty-two dishes, each 20 feet in diameter and 2½ stories high, scanning the sky, searching for any sign of radio emissions that may be from an intelligent civilization beyond Earth. The ATA is often used in stock video footage whenever UFO conspiracies arise in the news, or when the NASA Kepler program discovers new planets orbiting distant suns.

© The Author 2017

M. Brotherton (ed.), *Science Fiction by Scientists*, Science and Fiction,
DOI 10.1007/978-3-319-41102-6_10

Most of the people at the SETI Institute have PhDs and are actively involved in studying various areas of astrobiology — the study of life on Earth and beyond. The group involved with the ATA is very small in relation to our larger organization. It has dwindled to just a few people, eclipsed by the rest of the science being explored at SETI. When I started in 2008, I was working with a team of six. Now, eight years later, due to attrition and funding issues, I have inherited the entire job of observation at the ATA. That is bad for the SETI effort as a whole, great for me. I pretty much work alone. I keep the computers running, continually fix problems, writing code. A real geek's paradise. The most interesting part is that I am the guy searching for extraterrestrial life 365 nights a year. I decide what to observe, what frequencies to monitor and what to report. Sadly, though, I'm really the only one paying close attention. I comb through the data for signals with the help of computers. If there is a signal from ET received by the ATA, I am the guy who would have to determine if it is real, then decide what to do with that information. I think I am doing a good job, but I'm always my worst critic.

I work a lot from my home, in my garage, only occasionally going to the ATA. It is a long drive and most of the work can be performed remotely over the Internet. When computer hardware fails, I travel to the ATA, spending three to five days, over the weekend. If I go over a weekend or on a holiday, I am usually alone onsite. Just the 42 dishes, the computers and me. The dishes hum all the time, being cooled by air flowing through their cavities to avoid overheating of the receiving and motor pointing components. In the winter, if I get cold, I quickly warm up behind one of our racks full of computers churning away, searching for signals. The cooling fans push out a lot of heat. The place always feels alive, like something big may happen at any moment.

This is how the observation works: At any given time, the 42 dishes stare in unison at the same spot in the sky, tuned to look for radio signals at frequencies between 1 GHz (one billion cycles a second) and 10 GHz, usually focused on a single star. The faint radio signals arrive via underground fiber optic cables into our computer systems. The signals from all the dishes are converted from analog to digital and fed into a special device called a beamformer. The entire array of dishes is looking at an area of the sky that is 1/3 to 3 degrees in diameter, depending on the frequency being observed. The beamformer is a nifty device that further focuses attention on one really small point, removing the signal chaff coming in from that whole big area of the sky except from that small point. I can then focus on a particular star exclusively, cutting out most interference from other parts of the sky. Satellites and radio stations cause lots of trouble. They can potentially drown out signals of extraterrestrial life. As a consequence, signals received are immediately deemed suspicious, and most are identified as interference.

The thought is that if an alien sent a "Hello There" signal it would be sent at a frequency between 1 GHz and 10 GHz. This frequency range contains a lot less noise, or "static," than other frequency ranges; there is a natural dip in this range. As we figure, our best chances are within this frequency range.

Every night, I choose what to look at by dipping into a large star catalog for ideas. We like to concentrate on nearby stars and stars that are confirmed to have actual planets as identified by NASA's Kepler Mission. It really is a crap-shoot deciding what is best to observe. Occasionally, a new planet makes news because of some anomaly identified by the Kepler Mission data suggesting it is a good candidate for being habitable. We observe these planets soon after they are discovered and get some PR mileage out of it by writing blogs or short scientific papers.

I pay close attention to the observing, watching the system chunking away all night. I make sure nothing is impeding the observing, like a computer failure or one of the dishes having trouble pointing. Keep in mind that each dish is a big hunk of metal and it takes a lot of motor power to point them and keep them on target. I modified the software to detect problems and alert me by text in the middle of the night. What I really like to do is pay attention to the signal data. This is where we find extraterrestrial life amongst the cosmos static. The computers comb through the data creating statistics and reports. I analyze and tweak, telling myself I have to go to bed soon.

Needless to say this is a very interesting thing to do for a living, but constantly having something interesting prodding at me gets wearisome after a while. I wake every morning telling myself not to immediately log on and check out the previous night's results, not until I have my first cup of coffee and a "good morning" chat with my wife. But ET might be waiting.

***

OK, I won't keep you in suspense any longer. I saw *something*.

Well, I did not "see" it, it's more like I deduced it. Like I keep saying, it's complicated.

I started concentrating on a type of signal we call a transient. This is a signal that looks like just the kind of ET signals we want to see, but quickly disappear. We see it loud and clear, then five to ten minutes later it is gone. Why do they disappear? What are they? In most instances, we will never know, maybe it's a satellite or ground-based radio interference. We see a lot of transients, effectively disregarding them as interference. Several times I've gone back and looked for one of these signals and I have never been able to see it a second time. It's frustrating.

The Earth is rotating, so the ATA is either moving toward or away from a distant planet at any given time. Any ET signal received at the ATA will drift in frequency due to the Doppler Effect. The lack of signal frequency shift is the easiest way to determine received signals are simply interference. We see thousands of non-drifters every night that are deemed as interference. The observing system tries its best to weed though all the interference.

Like any other engineer, when I am not sure what to do, I write a computer program and have a couple more cups of coffee than usual. *Geek alert.* I love writing computer programs that can be automated to run daily. Although I am comfortable with several computer programing languages, my latest favorite programming language is Ruby. On the morning that I first saw a pattern in the transients, I had a working Ruby program ready in about an hour to comb the signal databases daily for all past transients and chart their frequency over time. The program could text the chart images to my phone before I got up in the morning. If ET was transmitting intermittently over time at the same frequency, I should be able to see some type of pattern in the graph of transients. A sort of sine wave should appear if I squint my eyes. I condensed the report down to five charts that would take about a minute to review every day. The thought was that over years of collecting data maybe would see something worth investigating.

The next morning I checked for messages on my phone. There were the charts. I sat hunched over in bed, pinching at my phone screen searching for the sine wave.

I saw a pattern. Before my morning coffee.

A mess of dots was on the charts. I thought my mind was playing tricks on me, but there it was, a pattern in the mess. A curvy sine wave.

It could be a coincidence, I thought. Random data points sometimes line up in a pattern by chance. I decided to look at the source of the signal every day for an hour and chart the transient signals, if there were any. Sure enough, every day like clockwork, the signal reared its head. Every 16 minutes and 31 seconds the signal would turn on and be loud and clear for five minutes and two seconds from the time it rose over the horizon till the time it set, only three hours a night. The signal was drifting in frequency, just as it would if it was emanating from a distant planet.

This was the perfect transient to avoid true detection by our software. I would never have seen it without manually looking at those charts. The signal detection scheme we routinely use looks in 92 second intervals, then stops, processing the data for signals, then looking again for 92 seconds more. This signal could easily be in the "off" state the next time or next several times our software looked at it. It is a tricky thing to know what to do with transient signals.

The signal appeared to be coming from a star in the Sagittarius constellation. This is the same constellation from which the famous Wow Signal emanated in 1972. But this star is far away from the Chi Sagittarii system. The frequency of the signal was in the 5 GHz to 6 GHz range. I won't be any more specific at this time, given the circumstances.

I bought some large capacity hard drives on my personal credit card so the gang at SETI would not get suspicious, and went to the ATA. I spent several nights collecting as much raw data as I could from one of the beamformers then made triple backups. Over the next several weeks, I continued to gather more evidence, more raw data.

***

Now what? I never actually thought I would actually see a signal that really panned out.

No one at the SETI Institute has ever briefed me about what to do if I ever see an actual ET signal. It seems strange. Maybe I missed that two minutes of employee orientation when I was hired. I always assumed I should tell my superior, but to tell the truth, I arrived at that assumption by reading science fiction books and watching movies.

"Contact" is one of my favorite movies and it was about SETI. In fact, before they made the movie some of the film crew spent time with the real SETI Institute employees to learn from them and be able to get the feeling for engineers so the actors would perform as close to real life as possible. They even took note of the pencils we used and what the coffee cups looked like. The protagonist of the movie, played by Jodie Foster, discovers an ET signal, admittedly much more interesting than the one I discovered, and look what happened to her! The government swooped in and pushed her aside, reducing her role to nothing. Would that happen to me?

The ATA does a good job at what it does, but admittedly there are radio dish sites that have hundreds of times the receiving capability of the ATA. Theirs just can't do surveys of large sky areas like we can. If the signal was made public, attention would quickly shift to one of the big guys, leaving me and the ATA in the dust. And, if that didn't happen and the ATA got more funding as a reward, they would be able to afford the best PhDs money could buy, and I'd be pushed out, just like in the movies.

What if I'm wrong about the signal? My lingering fear is that it's actually the result of a nearby farmer's old washing machine, or something even more embarrassing like a faulty baby monitor 20 miles away. I'm pretty sure of my pointing accuracy with the dishes. I regularly test the dish and beamformer pointing by trying to detect the Voyager1 signal. If I can detect Voyager1, which is outside of our solar system and has a transmitter the power of a refrigerator bulb, then the pointing is fine. I performed many tests by pointing off the ET signal location then back on again. That signal is definitely coming from the star I've identified.

But really, I should not be so paranoid. Can I think of some way to let people know about the otherworld signal in such a way as to benefit me and the ATA in the end? I know that if I cannot find any personal or organizational benefits, I still need to let the world know. How would I live with myself otherwise?

It's been nine weeks since I found the ET signal. Shouldn't I have reported this signal nine weeks ago? Shouldn't I report this now?

I decided to think like any other engineer with the desire to procrastinate. I wrote another computer program. I had all that raw beamformer data sitting on my desk at home virtually staring at me. I wanted to dig deeper into that data, hoping I could detect some pattern other than the timed 16 minutes and 31 seconds the signal is off then the five minutes and two seconds it is on. Maybe I could find a signal in the data. The computer that originally detected the signal has a generalized detection scheme that is not as high quality as it could be. It trades quality for speed, otherwise it would not be able to keep up with the data. There are less efficient, but higher quality ways to process the data. I thought I could see some structure in the signal. It was worth a shot.

The raw data from a beamformer and the data from the antennas, for that matter, is a series of power readings. In rough terms, this is basically representative of a sample of the signal power, 100 million times every second. These 100 million values are fed into our computers every second and transformed into frequency readings using a process called the Fourier Transform. The computer then searches these values for frequency values that slowly drift over time. This data is visualized as a waterfall graph.

Waterfall graphs look like static from an old TV, charting noise from the electronics, the environment, and random noise from space. If there is an alien signal, it will rise from the static and appear to have structure to it, like a straight slightly sloping line. The ET signal looks like a straight line that is shifting slightly in frequency. Actually it was somewhat curved, drifting slowly back and forth over the course of a day.

I had the luxury of being able to take my time processing the raw beamformer data, so I pulled out the really big guns and performed a multimillion-order Fourier Transform of the data. The result was much higher quality and took much longer than would normally be acceptable in our real-time signal searching process. The process created a really large waterfall plot and it took time to go through this data and find the signal. But there it was, drifting back and forth depending on whether the ATA was moving toward or away from it. I was able to zoom in.

The signal had little on/off sequences and looked to me sort of like a Morse code. Of course it was not Morse code, but on-off-on-on-off-off-on-on-on . I started thinking of it as dit-dit-dot-dit-dot-dit-dit. I was able to process a bit further and get the signal looking cleaner. Not great, but I could definitely see the on/off sequences in the waterfall. I had a lot of sick time saved up and for two weeks I hacked away at a program that was much more efficient and resulted in a readout of the dit-dot values. Eventually I had a display on my laptop that nicely showed the data. I was recording all the while, with backups of course!

I then spent another week making the program on my desktop work really well and look really cool. I made the most of my procrastination. I could not keep my eyes off the computer screen after that. Now the ET star was visible from 1-4 a.m., three hours a night I was receiving a signal from life beyond! I have no hope of deciphering it, but I wonder what the best artificial intelligence computer would make of it. Does the signal contain actual information we are supposed to interpret? Or is it just a hello signal? People smarter than I should be looking at this.

What should I do now? Nine weeks have passed. Staring at my computer screen makes me feel elated and a bit sick. Will I be remembered as the person who introduced ET to the world? Or, will I be remembered as that guy who *did not* tell the world? Should I put my head down, be a good engineer, and write another computer program?

This is no washing machine.

Afterword

SETI stands for the Search for ExtraTerrestrial Intelligence. The SETI Institute is based in Mountain View, California, employing more than 100 scientists, educators and support staff. Its mission is to explore, understand, and explain the origin and nature of life in the universe, and to apply the knowledge gained to inspire and guide present and future generations.

Founded in 1984, the SETI Institute initially purchased or was granted time on various radio telescopes to perform SETI radio signal searches. In 2001, Paul Allen (co-founder of Microsoft) agreed to fund the first phase of what is now called the Allen Telescope Array (ATA), which is located at the

Hat Creek Radio Observatory near Hat Creek, California. The array began observing in 2007. There are currently 42 receiving dishes, with the eventual goal of placing 350 dishes when funding is obtained.

The site has a small full time staff that keep the computers running and the dishes working. Maintenance and other activities take place during the day. The SETI radio signal search takes place during the night. The system was designed to be operated remotely, so the operator does not have to reside on site to perform the search.

The ATA does not have the ability to transmit radio signals, it can only receive radio signals.

The dishes scan the skies every night trying to detect radio signals that may originate from a distant civilization. At any given time, all the dishes point to the same location on the sky and the signals received from each dish are combined with the signals from the other dishes. Signal processing computers analyze the data for signals in real time, only keeping the small amount of data containing detected signals. As a result, most of the data received is discarded. Details of detected signals are stored in a database and raw data from the dishes pertaining to detected signals is stored on disk. Many gigabytes of data are gathered every night.

The receiving frequencies range from 1 GHz to 10 GHz, which is the range of frequencies used by satellites, radar, and spacecraft. As a result, there are a lot of radio signals detected every night and a large part of the search involves weeding out interference. This frequency range is very wide and far outside the range of human auditory capacity. Thus, we do not listen with headphones. The computers do all the work.

The data processed by the computers are most often represented as a graph called a waterfall graph. This graph looks like static on a TV. Any signals appear as patterns emerging from the static. The computers are basically picking out patterns from the static.

The ATA is designed to detect very weak and very narrow radio signals that drift over time due to the Doppler Effect. Signals that do not shift over time are categorized as radio frequency interference (RFI). Occasionally, signals of just the right type are seen and disappear. We call these transients. These may be cause by a satellite moving through the field of view, or an old transmitter malfunctioning.

Since initial operations in 2007, development and improvements have been a continual process. The signal searching schemes and data analysis algorithms are continually evolving. The antenna feeds in the dishes are being upgraded. New processing hardware and software is being developed. This results in the search becoming more efficient over time, increasing the chances of one day detecting a signal from ET.

As the operator of the SETI signal search, I often do fantasize about what it would be like if I detected a signal that was a good candidate. Within the SETI Institute, I have heard of several instances, before I came on board in 2008, where a good candidate was discovered. But in those cases the source was discovered to be coming from somewhere locally. So the search continues.

The Schrödinger Brat Paradox

Carl(ton) Frederick

Why, Roger wondered yet again, did a psychiatrist want him, a quantum physicist, to observe her patient. His curiosity, even stronger than his aversion to hospitals, had drawn him here to Mass. General to find out. *And, in any case, I can use a break from working on my theory.*

Now, gazing through a one-way mirror, he observed the psychiatrist — her welcoming demeanor and her warm smile. As for her patient, a clean-cut adolescent boy, he seemed completely normal — except that occasionally he conversed with someone who wasn't there. But Roger, judging from the end of the conversation he could hear, didn't think the unseen conversant urged the boy do anything cringe-worthy. *Like hearing one side of a phone call.* Roger shook his head, slowly. *Harmless.* He thought back seven or so years. His son at three had an imaginary playmate. *I think sometimes he still has one, but keeps him secret.* Roger looked hard at the patient. *This kid's a little old for it, but....* Roger took a deep breath and noticed that the room smelled vaguely antiseptic, a hospital smell. *I hate hospitals.* He transferred his gaze to the wall. The color was a calming baby-blue. Roger found it depressing. *I've got to get out of here.*

He returned his attention to the psychiatrist. He'd looked her up and found that she, Olivia Van Staaten, had a joint appointment in the university psychology department and also at Harvard's affiliate hospital. She had a reputation for innovation and had a good publication history.

Finally, the session ended and Roger walked briskly to the hospital cafeteria to meet with Dr. Van Staaten.

The cafeteria, Roger noticed, had a similar smell to the observation room — but this time tinged with the odor of bland, healthy food. He bought a cup of

© The Author 2017
M. Brotherton (ed.), *Science Fiction by Scientists*, Science and Fiction,
DOI 10.1007/978-3-319-41102-6_11

tea, carried it to a table, and sat, waiting. His hands wrapped around the cup provided a welcome warmth in a Boston January.

He stood when she came in and up to his table. She motioned for him to sit and they began to converse. Roger was pleased that she displayed the same warm personality that she'd shown with her patient. Almost immediately, they were on a first name basis.

"….and speaking of your patient…" Roger hesitated. "I felt a touch uncomfortable watching. Patient, Doctor confidentiality and all that."

"I commend your sense of privacy," said Olivia. "But the boy's parents approved. They said that if I thought you could help…." She spread her hands.

"But that's just it." Roger leaned forward. "How could I possibly help? Why me? I mean, I'm a quantum theorist, not a psychiatrist."

She looked down at her hands, and paused, as if about to take a leap into the unknown. "What do you know…," she said, tentatively. "What do you know about schizophrenia and Dissociative Identity Disorder?"

"All I know about it is," said Roger with a smile, trying to lighten the mood, "Roses are red. Violets are blue. I'm schizophrenic, and so am I." As soon as he'd said it, he realized it had been a bad idea.

Olivia's warm smile became noticeably cooler.

"Sorry," said Roger. "Australian kid humor. Not really appropriate here, is it?"

Olivia gave a slight shake of her head. "And not accurate, either. The verse should say DID, Dissociative Identity Disorder, not schizophrenia."

"Sorry," said Roger, again, "both as an apology and an expression of ignorance, "what is Dissociative Disorder, please?

" Dissociative *Identity* Disorder." She waited for a sign of understanding from Roger, didn't receive one, and continued. "Split personality."

"Oh."

"Actually," she said, distantly, "I don't really know what the patient's problem is." She seemed to be speaking to herself. "Similar to DID, but the personalities are very similar. Doesn't fit with classic schizophrenia either. And the usual psychotropic drugs, clozapine, risperidone, don't seem to have any effect." She pursed her lips. "And I don't really like having to prescribe them."

"If I might say," said Roger, "his…symptoms don't seem particularly severe. He seems to be a precocious, imaginative kid. Maybe that's all there is to it."

She nodded. "Yes, he is imaginative — too imaginative." She seemed to drift back into a conversation with herself. "Sometimes he imagines he's another person: a brother whom he doesn't have, a teacher, and even sometimes his own father."

Roger smiled. *That's very recursive.*

"And he has an eidetic memory," — She talked as if she were unburdening herself — "a virtually non-stop video recorder."

"I must confess I have that condition as well," said Roger. "My pose as an absent-minded professor is just that, a pose."

"Oh, I'm sorry. I didn't mean to imply that you — "

"No problems," said Roger with a laugh. "Mental issues are something of an occupational hazard with scientists, with great scientists at any rate. Newton, Tesla, Eddington, and even Dirac."

Olivia took a long sip of her coffee, seemingly hiding behind the cup.

Roger tapped the side of *his* cup and watched the waves propagate through the coffee. *Not that I can pretend to be great. But I'd willingly accept even more severe mental issues if I could get the wave-nature of my theory to work.* He bit his lip. *But I sincerely hope Bron hasn't inherited the issues from me.*

"Unprofessional as it may be," said Olivia, lowering her cup, "I'm becoming increasingly worried about him. The boy's a loner — not involved with his community. That's not healthy."

Roger made a sound to show he was listening.

Olivia seemed to take notice of him again. "Oh, and by the way, he's also expressed an interest in quantum physics. From what I can tell, he seems very knowledgeable about it." She stirred her coffee, absently. "We've done PET scans of the frontal cortex for the usual anomalies. A hint of an answer." She threw a glance to the ceiling. "I wish I could have used a higher intensity beam on the hippocampus."

Roger tried to look as if he knew what she was talking about — a skill he'd picked up in grad school. But she obviously saw right through him. "You *are* familiar with the hippocampus?"

"Um… the university for hippopotamuses?"

She sighed.

Sorry, said Roger, yet again. "More Australian humor, I'm afraid."

She offered up a tight smile. "They're in the medial temporal lobes of the brain. They're important for memory — and possibly for olfaction — smell."

"Very interesting," said Roger, suppressing his lack of interest. "But again. Why me?"

Olivia, looking uncomfortable, paused for a few seconds, then blurted out. "I read an article about the multi-world interpretation of quantum mechanics."

"Oh, dear." Roger gave a hint of a chuckle. "The Everett MWI theory."

"Yes. That was his name. You don't approve?"

"It's not that. Some prominent theorists believe it. A popular article, I presume."

"No. Not really." Olivia sounded defensive. "Semi-popular, perhaps." With her forefinger, she traced out lines on the table as if reading from them. "The article said that multiple universes communicate weakly through things called interference and superposition. Is that more or less correct?"

"Actually, yes. But what does that have to do with...." He felt his eyes go wide. "Are you.... Are you suggesting that your patient is actually communicating with himself in another universe?"

"I've explored every other solution I could think of and, as Sherlock Holmes says, once you've eliminated the usual possibilities, whatever is left...." She spread her hands. "*Is* it a possibility?"

"I don't really know. But it does seem maybe a touch farfetched."

She stared intently at him. "*Too* farfetched."

Roger, wilting under the intensity, was reluctant to dissuade her. "Maybe not. But why are you so obviously keen on it?"

She gave a hint of a laugh. "Haven't you ever become obsessed with your work?"

Roger laughed. "Guilty. I've been struggling virtually round the clock to derive the Schrödinger equation from my own theory, Indeterminate Granular Space-time Mechanics." He chuckled again. "It seems we're kindred spirits, you and I."

She raised her coffee cup, as if in a toast, and said. "The Schrödinger equation. The article talked about that and also the Schrödinger Cat Paradox — and something called the wave function, and said it was crucial for multi-worlds. But I must confess I didn't fully understand it."

"In the best physicist tradition," said Roger as he pulled out a napkin from the dispenser and laid it flat on the table, "we'll use this as a blackboard." He extracted a pen from his shirt pocket and wrote an equation. "This is the Schrödinger equation." He pointed to the Greek letter Ψ. "And this is the wave function."

Olivia stared at the napkin "I can't say that this adds much to my understanding. But actually it's not the Schrödinger equation that I'm interested in."

Roger didn't reply as, staring at the equation (which was unnecessary as it was long etched into his being) he got an idea. *Maybe the equation really describes fluctuations of the metric tensor.*

After some long, silent seconds, Olivia made a throat clearing sound.

It brought Roger out of his physics world.

"Oh. Sorry. Um..as for the Cat Paradox...."

"Oh. I think I understand *that*." Olivia looked up from the napkin. "If a quantum experiment can produce a dead or a live cat, it provides a combination of both until someone actually checks on the condition of the cat."

"Yes. A *superposition* of a live and dead cat. And only when someone checks, will the cat be alive or dead."

Olivia frowned. "I must say I found that ridiculous."

"Schrödinger thought so, too. But that's what his equation seems to predict. MWI, on the other hand says that there will be both a dead cat and a living cat, but in different universes."

"That sounds ridiculous as well."

Roger nodded.

"But," Olivia went on, "the article said there may be some communication between universes — especially in cases where the subjects are not tightly connected to the rest of the world."

"Which," said Roger, "is why the cat in the Cat Paradox experiment is placed in a sealed room."

"Experiment? Do you think an experiment is really possible?"

"Possibly." Roger felt it best not to mention that his IGSM was a counter theory to the Multi World Interpretation. *And if an experiment were to cast doubt on MWI, I would not be unhappy.* He was surprised at himself; He was actually beginning to consider an experiment. *And not just a gedanken experiment.*

Olivia got a faraway look. "I'm sure the university's animal research committee would not go along with killing a cat. And in any case, a brain scan on a dead cat wouldn't prove anything."

"We can alter the experiment." Roger smiled at the thought. "We could use a white cat...so...so that instead of the cat dying, we'll dye it. We'll have it dropped into a vat of black hair dye. Then we'll have a superposition of a white and a grey cat."

"I assume," said Olivia, "that a...a superposition of a white and black cat isn't a grey cat then."

"No."

Olivia nodded. "I wonder. My patient. Perhaps I should have treated him in your...cat box, .a box isolated from the rest of the world." She hesitated. "Although I'm afraid he's already isolated from the world. Already a recluse."

Roger gave a long sigh. "I've been something of a recluse myself since my wife died."

"Oh I'm sorry."

"About four years ago. A neurological disease."

Olivia gave a nod of understanding, and stayed silent.

"I felt I needed a complete change," Roger went on. "I left The University of Sydney for a post here at Harvard. I threw myself into my research. I think about it all the time, my research I mean. I dream about it. And I'm afraid I'm short-changing Bron, my son, by living in my research world more than in his." He flashed a smile. "Sounds sort of like MWI, doesn't it?" He forced a chuckle and glanced back to the interior of the hospital. "In a way, I might not be all that different from your patient back there." He leaned back in his chair. "But I'm talking too much. Apologies."

"No, no. Go on. You're doing me a real favor by coming here."

"Not at all. In fact you're doing me the favor. I've become very interested in the workings of the brain ever since...."

Olivia placed a hand over his.

"I've been watching my son for any early signs of neurological problems."

"And?"

Roger forced a laugh. "Who knows? He's ten. Hard to say what's normal. He's in a counting things phase: number of pairs of underwear he has, number of game cards, number of steps from home to school."

"That's normal for ten."

"He does seem most of the time to be living in multiple worlds — ever since his imaginary playmate years. Very convincing playmate. Sometimes I found myself believing in him as well."

"I wonder," said Olivia, thoughtfully, "if people, lonely people not well connected to the community, who show symptoms of DID actually are evidencing multiple universes."

"And ditto, young kids."

She nodded. "Maybe them, too."

"I'm the quantum theorist," said Roger, lightly. "I'm supposed to be the one with the nutty ideas."

"You're saying that the idea is nuts?"

"Well…" — Roger thought about Bron when he was a toddler. — "Maybe it's not nuts."

"Is it perhaps subject to experiment?" said Olivia. "DID patients do seem to have identifiers in the cortex. But CAT scans shouldn't be done too often on people — especially the high intensity devices we need to use. But maybe we could do the scans on...."

Roger smiled. "On Schizophrenic cats?"

"On DID cats."

In his mind's eye, Roger observed the family cat strolling serenely to its food dish, and then running madly from room to room as if being pursued by the hounds of hell. *MWI?* "Maybe we could perform an experiment."

"Do you believe in MWI?" said Olivia, abruptly.

"Me?" Roger struggled for an answer. "I...I do, sort of, sometimes."

Olivia laughed. "You sound sort of indecisive."

"I used to be indecisive," said Roger, with a smile. "but now I'm not so sure." *Damn. Why am I doing this? She wants me to be serious.* "My belief is compartmentalized. In the context of quantum physics, yes, I consider it possible, *barely*. But as far as the real world..." Roger shrugged. He pondered for a moment. "But beliefs aside, I think I can come up with an experiment."

"Soon?"

"A few weeks. A month, maybe,"

***

The Schrödinger Cat Room, a room within a room, dominated Lab 202 in Harvard's Physical Sciences building. Raised five inches off the laboratory floor by pneumatic dampeners, cocooned within batts of sound absorbing insulation, and with self-contained battery powered lights and air pumps, the Cat Room could be effectively isolated from the rest of the universe. The exception was an umbilical cord supplying power which would be unplugged prior to running the experiment.

The whitewashed starkness of the room was relieved by text inscribed in blue on an outside wall, painted by Dexter, Roger's grad student. He, avowedly nonreligious, had adorned the door with the somewhat devotional inscription,

$E = h\nu$...and there *was* light.

The room had a small door and no windows. The door, at the moment, stood ajar.

Inside the room, Roger struggled to get the cat-drop module to work while outside, Bron sat struggling to get to level eight in his video game.

External to the Cat Room, sat a six-by-six foot cage, the Cat Corral, containing two automatic litter boxes, four filled food and water dishes, and four towel-lined cardboard boxes to serve as temporary housing. At the moment though, the Cat Corral was devoid of cats.

The lab, lit by banks of fluorescent lights, gave the room a shadow-free brightness and a hint of fluorescent tube hum.

"Damn it to bloody hell!" Roger, clutching the screwdriver like a lethal weapon, stormed out of the Cat Room, back into the lab.

Bron looked up from his game and stared at him with accusing eyes.

"What?" said Roger.

"You yell at me for swearing. It's not fair that you're allowed to swear."

As Roger tried to frame an answer, Dexter flew into the lab. "Hi, Dr. Tate." Then he noticed Bron. "Hey, Brat." He laughed. "Shouldn't you be in school now?"

Bron waved, hi.

"Bron was suspended for the day," said Roger, "for coming to school armed."

Dexter cocked his head. "Armed?"

"A water pistol."

"Oh."

Bron smiled, sheepishly.

"Water pistol!" said Roger "That's an idea."

"What's an idea?" said Dexter.

Roger explained his problem with the cat drop, then turned to his son.

"Bron. How would you like to earn another game cartridge?"

"That would be… nice," said Bron, warily. "How?"

"The cat-drop mechanism doesn't work, but the photon gun and the beam-splitter, the quantum coin flip, does." Roger bubbled with enthusiasm. "So we'll replace the drop mechanism with you."

"Huh?"

"So you fill your water pistol with the black hair dye. You'll run the coin-flip. I'll show you how. And if the flip-LED flashes, say green, you do nothing. And if it flashes red, you shoot the cat until it's all black."

"Fun!" Bron paused. "But I want *two* game cartridges."

"This is extortion," said Roger with a laugh. "But, it's a deal." He held out a hand and Bron shook it.

"Hey, wait a minute," said Dexter. "It's a different experiment, now."

"How so?" said Roger.

"Well, originally we had a superposition of a live and dead cat — "

"A white or black cat," Bron interjected.

"Yeah, Fine." Dexter went on, "So the cat wouldn't have a well-defined… color until we open the box and look."

Roger nodded.

"But now," said Dexter, "we'd also have a superposition of Brats — of one who painted the cat and the one who didn't."

"Hmm." Roger considered it. "But Bron isn't the subject of any quantum measurement." He paused. "I'll show you how to think on that." He blew out a breath. "I really don't actually believe in MWI. But… but then again maybe we shouldn't run the experiment." He looked down at Bron. "What am I saying? He's just going in to possibly paint a cat."

"I'd be happy to paint the cat," said Dexter.

Bron protested. "That's my job. It's worth two game cartridges."

Roger sighed. "No, Dex. I need you to baby the CAT scanner."

"Yeah, fine," said Dexter. "And anyway, I don't believe we'll have multiple Brats. I can't really believe that any time Brat might or might not sneeze, we get a superposition of Brats."

"No. Not likely." Roger gave a snort of a laugh. "I was just being silly."

Dexter glanced over at the Corral. "But where are the cats?"

"Olivia's bringing them — courtesy of the Tufts Veterinary School." Roger checked his watch. "Should be here by now." Almost by reflex, he glanced out the window. "Ah, a cab's pulling in up front." He watched for a few seconds. "Yes, that's her." He turned to his son. "Bron, would you go down and help Doctor Van Staaten bring in the cats?"

"Sure." Bron sprinted out of the lab.

Roger continued staring out the window for a few moments. "Rotten weather. Looks like a bad storm's coming." Just then came a peal of distant thunder.

"All right," said Roger, turning from the window, "let's get set for a run."

"Right!" said Dexter.

While Dexter ran a calibration on the CAT scan unit, Roger disconnected the umbilical, then darted into the cat room to turn on the photon gun and beam-splitter unit. As he came back into the lab, Olivia and Bron walked in, each carrying two occupied cat carriers.

Dexter ran to help. "I'll take those, Brat." He relieved Bron of the carriers and sprang to the corral.

Olivia, though, frowned at Dexter in obvious distaste. "Brat?" she said. "That's not a nice thing to call someone."

Dexter looked confused as to how he should respond, but Roger just laughed. "Brat," he said, looking at Olivia. "Bron Rattray Arthur Tate... BRAT." He walked to be beside his son.

"Oh, for crying out loud." Olivia smiled but it seemed forced. After a pause, she said. "Why not Bron Arthur Rattray Tate...BART?"

"In another universe, perhaps," said Roger, lightly. "We flipped a coin when he was born. And anyway, my middle name is Allen — which makes me RAT." He scrunched his shoulders. "Rat, Brat. Seemed fitting, somehow."

As they talked, Dexter transferred the cats, each one as white as a full moon in winter, to the corral. Each cat wore a blue collar with a transparent pocket displaying the cat's identification number.

Olivia looked from Roger to Bron. "Bron Arthur Rattray Tate. Such a long name for a young lad."

"An Anglo-Australian custom," said Roger, tousling his son's hair. "especially if there's a number of wealthy relatives to keep happy."

Bron wasn't paying attention. He was watching Dexter with the cats. "They're beautiful," he said as if to himself. "Green eyes, long whiskers, big ears, and big furry paws."

Roger followed Bron's gaze. "All right, Dex. Bring me a cat. Let's get going with the experiment."

As Dexter brought over a cat, Bron asked, "Can I hold him?"

"Sure," said Roger.

Dexter transferred the Cat to Bron.

Roger went into the Cat Room and urged Bron-cum-cat to go in as well. He took the cat and placed it in a little cage suspended over the vat of quick-drying, black hair dye. The bottom of the cage was a trap door. "The cat is supposed to fall into the vat if the beam-splitter collapses the photon to the detector."

Bron fidgeted. "Do we have to drop the cat into the dye?"

"I wish we could but I can't get the mechanism to work. So you'll have to... to paint the cat."

Bron fidgeted some more.

"So fill your water gun from the vat."

Bron did so.

"Okay," said Roger. "So here's how it works. When I leave and close the door, count to, say, fifty." He pointed to a button on an instrument console. "Then push this button. It triggers the photon gun" He pointed to an LED over the button. "If this lights up red, shoot the cat until it's completely black. Then open the cage and let the cat loose. In about ten minutes from now, I'll open the door and you and the cat can come out. So far, so good?"

"Yeah," said Bron, tentatively, dividing his attention between his father and the cat.

"And if the light is green instead of red, don't shoot the cat. Okay?"

"I guess so."

"Bron. Pay attention."

"I *am* paying attention."

"Don't worry." Roger glanced at the cat. " This won't hurt the cat at all." *But we'll have one very angry wet cat.*

"Yeah, fine." Bron 's voice held a trace of trepidation.

"Good." Roger turned. "Oh. And don't try to come out. I'll let you out. The door will be locked to make sure the Cat Room is truly isolated from the rest of the world."

"I don't like the idea of being locked in."

Roger patted his son the head. "Only for ten minutes. I'm sure you can stand that. Yes?"

Bron nodded.

Roger left the Cat Room. He gave a thumbs up to Bron, then closed and locked the door.

He looked over at Dexter at the CAT scan unit. "Every thing okay?"

"Perfect," said Dexter.

"Good." Roger gestured at the wall clock. "In ten minutes, we'll see." He stared at the clock — and continued to stare at it. He did so as a cover, so he could think about his physics theory without being expected to engage in conversation.

Olivia and Dexter exchanged glances, but kept silent.

Roger bit his lower lip and thought, deep. *Okay. I know the particle must oscillate at its Compton frequency. But why and how? What is the mechanism. It must have something to do with time.* His eyes on the clock, Roger pondered time — until the minute hand indicated that ten minutes had elapsed. He broke from his thoughts and returned his attention to those around him.

"Okay, here we go." Roger turned, unlocked the Cat Room Door, and pulled it open.

The cat, still white, ran out. Bron followed, slowly, and avoiding eye contact with the others.

With Roger and Olivia watching, expectantly, Dexter grabbed the cat and petted it until it became calm.

"All right, cat," said Dexter, "Now for a nice little CAT scan." He strapped the cat onto the bed of the scanner.

After a few minutes, Dexter looked away from the scan console and shook his head. "Negative."

Olivia cast her eyes down.

Roger let out a heavy sigh. "It may be that MWI is simply wrong."

"We could try it again with another cat,' said Dexter.

Roger gave a non-committal "Hmm."

Olivia and Roger watched in silence then as Dexter unstrapped the cat and returned it to the corral.

Bron pawed the ground, nervously. His sneaker breaking the silence by making high-pitched squeaks on the tiled floor.

The sound drew Dexter's attention.

"Brat. You did carry out your part, didn't you?"

Bron continued pawing the ground. "Yes.... Sort of.... Not exactly."

"What?" said Roger, sharply. He swiveled to face his son. "What do you mean?"

"I didn't really understand what you wanted me to do."

"Why didn't you ask me?"

"I don't know."

Roger threw a quick glance of exasperation upward, then returned his gaze to Bron. "Then, what *did* you do?"

"I counted the floor squares — "

"The floor tiles?"

"Yeah. If there was an even number of floor tiles, I'd paint the cat. If there was an odd number, I wouldn't."

"That means there was no quantum experiment," said Dexter. "So, of course the experiment failed."

"Because Bron used the odd or even floor tile number to paint or not paint the cat," said Roger, softly, as if to himself, "you're saying there was no quantum universe splitting?"

"Yes. Exactly," said Dexter. "Of course."

"I'm not so sure."

"But the number of squares is known — "

"But not by Bron," said Roger.

"Still," Dexter persisted, "it's a number, deterministic, not stochastic."

"If true stochasticity were required," said Roger, "then that would rule out the DeBroglie-Bohm picture." *And my IGSM model as well.*

Without comment, Olivia watched the interchange as if a spectator at a tennis match.

Clearly wanting to avoid a theoretical physics confrontation with his professor, Dexter diplomatically changed tack. "You know," he said, "Considering the Cat Room floor dimensions, I'd imagine the room would most likely have an even number of tiles. Odd times odd would give odd, but everything else would give even. "He turned to Bron. "By the way, Bron, how many tiles were there?"

"Hundred seventy six."

"Ah. Oh, wait." Dexter looked over at the Cat Corral. "One seventy six. Then…then the cat should be black." He looked accusingly back at Bron.

"Well…um…," said Bron, wilting under Dexter's gaze.

Roger stepped in to save his son. "Okay," he said. "We'll do the experiment again. This time, we'll practice it, first. Bron" — Roger pointed — "into the Cat Room."

Bron, obviously happy to get out from under Dexter's accusing gaze, sprinted into the Cat Room.

Roger turned to his grad student. "All right, Dex. Hand me another cat." He paused. "No, wait. I'll get him after we practice."

Roger sprang up into the Cat Room and let the door swing closed behind him. He turned to Bron. "All right then. Why *wasn't* the cat black?"

"Well, I did count the squares." Brat retreated, defensively, to a corner of the room. "And there were really a hundred seventy six of 'em, but…but I decided I liked the cat being white."

"What?" Roger slapped his hand down on the console, triggering a quantum decision — red. "Bron Rattray Arthur Tate. I should ground you for a *year*." He let out a breath. "In another universe, Bron, you're now probably getting the spanking of your life." *Why did I say that?*

Bron looked horrified and absently moved a hand to cover his bottom.

"Hey," said Roger. "I was only kidding. You know I never hit you. *Although I've been sore tempted on occasion.* He moved to his son, gave him a fatherly hug, then said, "Let's get out of here." He moved to the door, but it wouldn't open.

Damned lock system. Roger pounded on the door, knowing that it was likely useless. The room was designed to be an isolated unit. But still he pounded. There was nothing else he could think of to do.

After a few minutes where Bron's expression morphed from amusement to concern to fear and then to terror, the door opened.

"Sorry," said, Dexter. "When the door swung closed on its own, the lock must have engaged. I didn't notice it until just now."

"No worries." Roger bit his lip. "But…but something strange occurred in there." He described what happened.

"Hey," said Dexter. "Maybe he *was* experiencing a spanking — in another universe. You might have caused a quantum decision. And…and because the room was so detached from our universe, you were able to sense a little bit what was going on in an alternate one." He looked at Bron. "It that it, Brat?"

"I don't know."

"But my saying Bron was being spanked in some sense caused it to happen." Roger rubbed a hand across his forehead. "And that would mean human thought affects quantum mechanics."

"I read that Wigner and DeWitt believed that," said Dexter.

Roger nodded. "They did, indeed."

Olivia spoke up. "It's never appropriate to hit your child."

"What?" Roger swiveled to look at her. "I don't. Never have."

"In another universe, you do."

"Sorry," said Roger, not sure for whom he was apologizing.

"No. *I'm* sorry," said Olivia with a contrite smile. "I sort of assumed that you and your alter ego used the same methods of discipline. I guess I was trying to chastise your other self."

Roger chuckled and pointed at the Cat Room. "I think you would have had to do it in there."

"Speaking of *in there*," said Dexter. "Aren't we going to run the experiment again, today?"

"Right," said Roger, firmly. "We'll run it again, now." He glowered at Bron. "You'll cooperate. Yes?"

Bron nodded. "Yes," he said, weakly. "But…but we didn't practice."

"Okay, okay." Roger, suddenly impatient to just run the experiment, let out a sigh. "We'll practice, first. And this time we'll do it with a cat." He turned to his grad student. "Dex, a cat, please."

"You're sure it'll work, now?" said Olivia.

"Am I sure? Strewth, mate. I'm a quantum physicist." Again, Roger chuckled. "I'm not certain about anything." He pointed to a wall poster. "Dexter put that up last week."

Blessed are the uncertain, for maybe they shall see Heisenberg!

Olivia looked away at the sign. Her face registered incomprehension.

Dexter looked sheepish.

There came a flash of lightning, and the lights flickered.

"Nothing to worry about," said Roger. "Happens all the time." As he said it, there came the sound of heavy rain.

Dexter threw a quick glance to the window, then fetched a cat and handed it to Roger.

"Okay, Bron," — Roger accepted the cat and nodded a thank you — "let's go!" Roger and Bron entered the Cat Room. As Roger moved to close the door, a second peal of thunder came, this time reverberating through the lab — and the lights went out.

Still holding the cat, Roger opened the door fully and looked out, hesitating on the raised threshold of the Cat Room. The battery-driven lighting from its interior cast black silhouettes of him and the cat against the laboratory floor made grey by the storm-cloud darkened illumination from the windows.

"Now that we're off the power grid," said Dexter, softly against the darkness, "we have yet another degree of isolation from the rest of the world."

Roger didn't know if he should join the others in the lab or retreat to the fully-functional and illuminated cat room.

Dexter seemed to read his mind. "I expect the power to come on within a minute or so. You might as well go ahead with Brat's practice session."

"I agree," said Olivia. "I'm very eager for another trial run."

Roger nodded. "Dex," he said, "How long will it take to restart the scanner once we get power again?"

"Just a couple of minutes."

"Okay. Fine. See you soon." Roger turned and entered the Cat Room.

Inside, Roger gingerly checked if the door lock was disengaged. It was fine. He placed the cat in the cat-drop cage, then turned to Bron. "Ready?"

"Yeah." Bron pulled out his water pistol and aimed it at the cat. "Pow!"

"Not now," said Roger. He guided Bron to the control console.

"Okay. Like last time, pushing this button fires the photon gun." He lifted Bron's hand by the wrist — the hand not brandishing the water pistol, and placed it on the console. "Now push the button, please."

Bron did so and the LED above the button went green.

Roger pointed to it. "All right, what does green mean?"

Bron shrugged. "I don't know."

Roger tried not to shout. "It means *don't* paint the cat! Push it again."

Bron looked up at his father. "Is it true that whenever I push this button, I'm creating a new universe?"

"So they say."

"Wow. Then I can create a real lot of universes. Ping! Here comes another one." He pushed the button.

The LED shone red.

"And what does this mean?"

"Um…that I *should* paint the cat?"

Roger raised a fist. "Yes! Now let's practice it some more."

"Why only in here?" Bron pointed his water pistol at the door. "I mean, why can't I create universes out there?"

"Our universe is kept from splitting by continuous measurements."

Bron wrinkled his nose. "Huh."

"Out there," said Roger, gesturing toward the door, "everything is in some way connected to something else. But we're in a sealed box, unconnected to out there. So here, we can create universes. Push the button, please." *I'm almost beginning to believe this, myself.*

"Okay, okay."

Roger imposed a dozen or so repeats of the button pushing and interpretations, and then said, "All right. I think that's enough."

"I'll say, it's enough."

Roger went to the door and turned the knob. "Oh, not again."

"What's the matter?"

Roger forced a smile, not wanting to upset Bron. "I'm afraid we're locked in again. But don't worry. I'm sure Dexter will notice soon, and will let us out."

"I'm not worried," said Bron.

Roger didn't believe him, and to keep him from panicking as had happened the last time, he had Bron go back and practice some more. *Grumbling is better than panicking.*

After many minutes of practice, Bron said, calmly, "It's taking Dexter a long time."

"Yes, it is." Roger had tried to keep his voice even, but he was growing impatient — and a little concerned, concerned mainly by the fact that Bron, who hated the idea of being locked in a room, *didn't* seem concerned. *Especially since he was scared out of his mind the last time. Kids are strange. Olivia was right; kids are not completely bound to our world.* Roger glanced at the cat in the cage. *And neither are cats.*

Bron, now almost by reflex, pushed the button, and the LED glowed red.

"Can I paint the cat, now?" Bron pleaded.

"No."

"Why not?" said Bron. "We can do the experiment ourselves."

Roger shook his head. "It's not an experiment unless we can learn something from it. We'll have to CAT scan the cat." For emphasis, he pointed to the cat cage where the cat was struggling to break free. "We might as well let him out."

"Great!" Bron went and freed the cat.

The cat meowed, loudly, and ran madly around the room, stopping finally at the door.

"I wonder how he knows it's a door," said Bron watching the cat pawing at the entrance.

Roger, staring more at the door than at the cat, gave an 'I don't know' shrug.

"I'm bored," said Bron. "I'm going to count floor tiles."

Roger, boredom mixing with concern, watched him count.

"That's funny," said Bron, at length. "I got one seventy two this time. Wonder where I messed up."

The cat meowed more loudly while again pawing at the door.

Just then, the lock clicked.

Roger rushed to the door to open it, but before he could get there, the door itself opened, revealing Dexter, framed by bright, fluorescent lighting.

Roger squinted against the light. Clearly the power failure had been fixed. He let out a breath. Despite himself, he felt a sense of relief. The cat though, ran not out the door but further into the Cat Room, stopping under the control console.

"Ah," said Dexter, throwing open the door full wide, "I thought you might be in there."

Roger, puzzled by the comment, canted his head.

As Dexter stepped aside, Bron and Roger left the Cat Room for the expanse of the laboratory.

But Roger's head remained canted in puzzlement; Things looked different, somehow. And Olivia wasn't there.

"Oh," said Dexter, "You're wondering why I'm here." Dexter explained that the stipend committee meeting had been canceled because of the snow storm.

"What meeting?" said Roger, with narrowed eyes.

"The Stipend Committee. I told you about it."

"You did?" *And what snow storm?* Out of the corner of his eye, he saw snow on the window sill. *Something's very wrong here.*

Dexter nodded, then turned his attention to Bron. "Hi, Bart," he said, cheerfully. "Think you can handle the experiment now?"

Bart? Roger jerked his head to look wide-eyed at Bron. *Here he's called Bart...and he's okay with that.* Roger stiffled a breath. *HE did it. HIS mind determined this universe.*

"Hi, Dexxy. Yeah. I can handle it."

Roger wondered if perhaps he should see Olivia, professionally.

As he wondered, Roger saw a black cat saunter out of the Cat Room. His mind overflowing with a kaleidoscopic welter of superimposed images, Roger watched idly as the white cat padded to the laboratory's far window, then leaped to the window sill and looked out onto the world — its black fur stark against the pristine snow.

After a few seconds, his mind having partially cleared, Roger ran to the window, scooped up the cat and shouted to Bron. "Quick. Back into the experiment room."

"Are you going to hit me again?"

Roger braked to a stop. "Am I going to *what*? No. Of course not. Get into the room!"

"Why?"

"Don't argue. Go into the room."

Roger, at a run, backtracked the trail of steadily fading black paw-prints toward the Cat Room. Then, still holding the cat, he bundled Bron into the room.

"Why?" said Bron, again, this time louder.

"To count floor tiles."

"I've done that."

"You'll count them again and again."

"Huh?"

"Until you get one hundred seventy six."

"Huh," Bron repeated.

Roger pulled the door closed. He heard the lock snap.

Roger dropped the cat and slumped back against the door. The Cat Room felt now like a refuge.

"What's going on," said Bron, his voice an accusation.

"Dexter, out there," said Bron. "He called you Bart."

Bron wrinkled his nose. "No he didn't."

"Please, Bron," said Roger recognizing his pleading tone but unable to do anything about it, "humor me and count the tiles."

"Four game modules?"

Despite himself, Roger laughed. "Yes. Deal." *It's the same Bron — my Bron.*

Bron got to his knees and began counting. It was a slow process.

Roger, watching, retreated again to his physics. "If only I could justify the Compton Frequency," he said softly. *You might think about the space Wiener-process flip rate being different from the time rate.*

"What?" said Bron, looking up.

"Just physics mumbling. Ignore me."

Roger nodded. "And if the time rate were much longer than the space rate, that could generate something like an oscillation." *Yes. I think we've got something.*

Again, Bron looked up, but only briefly before continuing his counting.

Roger lost himself in his work until Bron said, "That's funny. This time I counted one seventy three." He sounded genuinely puzzled.

"Count them again."

"Yeah, I think....." Bron sounded as if he were talking to himself. "I think I will."

He did, and came up with a still different number of tiles.

Over time, Roger, preoccupied with quantum theory, lost track of the number of Bron's counts.

"The ratio of the rates is a function of the mass. Yes?" Roger slapped a hand over his fist. "This must be it." *Yes. You're right. I'm sure we've got it. And I think it will generate a value for the Planck mass.*

Roger blew out a big breath and smiled. He'd been working on the problem practically forever. "Roses are red," he said under his breath. *Violets are blue.*

"One seventy six," Bron proclaimed.

"What? Great!" Roger jumped to the door. The lock clicked as he reached it. He pushed and the door opened. The lights were still off in the lab, and there was the sound of rain.

Roger felt the cat brush against a pant leg as it ran into the darkness.

He saw Dexter starting after it.

"Dex," Roger called out. "Forget the cat. Come here, please."

Dexter padded over. "Yes?"

Roger pointed to Bron who'd come out of the Cat Room and had hopped down to the lab floor. "Dexter, what's his name?"

"Excuse me?"

"What's his name?" Roger insisted, "his full name."

Dexter, looking puzzled, said, "You mean Brat? It's Bron,…uh Rattray I think, Arth — "

"Good, wonderful. Thank you."

"Roger, what's wrong?" came Olivia's voice.

"Oh, I am so glad to see you," said Roger.

"Why now, especially?" She seemed truly puzzled. "Did something happen in there I should know about?"

Roger made the snap decision not to tell them about the other universe. They'd think he was crazy. Instead, he said. "Great news. I've solved the main problem with my theory."

"Hey," said Dexter. "Congratulations."

"Thanks. And as they say, two heads are better than one."

Dexter wrinkled his nose in clear puzzlement.

Olivia narrowed her eyes and said, softly, "Why do you say that?"

"Yeah. Why *did* I say that?" said Roger with a chuckle. *Because it's true.*

With a hum, the lights in the lab came back on.

"Power!" said Dexter. He glanced at Olivia and then Roger. "Ready for another experiment run?"

"No!" said Roger, almost at a shout. Then, feeling sheepish, he said, "I'd like to wait until I get the cat-drop mechanism working. Then we can run the experiment properly."

Dexter and Olivia stared at him, expectantly.

Roger merely smiled. *I'll tell them the details later, when I've figured them out myself. Right now, they'd probably think I'm crazy.*

Under her professional smile, Dr. Van Staaten suppressed a sigh. It had been an unproductive session with her young patient. He scarcely paid attention and his mind wandered and raced like a motor boat with a broken rudder. It was not a good way to start the year.

She stood and ushered the boy out of the treatment room with its baby-blue walls that she found so depressing. "We'll send you back to the play-room for a while, Bron — while I have coffee and a chat with your father."

Afterword

Why, I've long wondered, does quantum mechanics work the way it does.

Although it is a remarkably reliable schema for describing phenomena in the small, quantum mechanics (QM) has conceptual problems; e.g. How can 'entanglement' send information faster than light (and without violating relativity)? How can it be that the wave function Ψ (psi) can instantaneously collapse? In what medium does Ψ travel? What is Ψ? What explains superposition? What is happening in the two-slit experiment? Can the two-slit experiment (at least in theory) be performed with macroscopic masses? Is 'The Cat' alive or dead? — As a theoretical physicist, this is the sand box in which I play.

In the above, I've given prominence to the 'two-slit experiment' since Richard Feynman said that "all of QM can be understood by understanding the two-slit experiment." But then he added, "But unfortunately, no one understands the 2-slit experiment." To be fair, he also said he didn't know if there really was a problem with QM — but if there was, it was a big one.

The problems surrounding QM center on the Schrödinger Equation (the master equation of QM) for a "particle,"

$$ i\hbar \frac{\partial \Psi}{\partial t} = -\frac{\hbar^2}{2m}\frac{\partial^2 \Psi}{\partial x^2} + V(x)\Psi, $$

defining Ψ which encapsulates everything that can be known about the particle. [$V(x)$ is a potential, akin to potential energy, and $\hbar$ is Planck's constant divided by 2π.] The square of Ψ is the probability of finding the particle at a particular point in space and time. But what exactly is Ψ? There are two main schools of thought:

1 - The Copenhagen Interpretation (of Niels Bohr) and the related MultiWorld Interpretation [1], MWI, (of Hugh Everett). In the Copenhagen view, Ψ *is* the particle. The particle doesn't have a trajectory: i.e. its momentum and position do not exist simultaneously. Ψ collapses to a point (instantly, throughout space and time) when the particle is measured. Copenhagen is the dominant paradigm for QM. Most of us quantum theorists have 'grown up' with it. But recent work on 'weak measurements'[2] has thrown Copenhagen into doubt by strongly implying that quantum particles *do* actually have trajectories.

The MWI is the same as Copenhagen except there is no wave function collapse upon measurement. Instead, all possible measurement results occur — but each in a separate universe. That is to say that any quantum measurement (that can have multiple values) creates one or more universes. A number of good theorists believe this. But Feynman called it "nonsense." [Even though I used MWI in my story, I nonetheless agree with Feynman that it is nonsense — as well as a profligate waste of universes.]

2 - The Ghost-wave/Hidden Variable Interpretation [3] (of Count Louis de Broglie and David Bohm), the Stochastic Mechanics [4] (of Edward Nelson), and the Indeterminate Granular Space-time Mechanics [5, 6] (of C. Frederick [me]).

In the de Broglie-Bohm Interpretation, Ψ is, as de Broglie says, 'the ghost wave that guides the particle' (but what that wave *is* is not explained). Particles do have trajectories, determined by 'hidden variables'. The mathematician, John Bell, though, showed that for that to be possible, the theory must be 'non-local' [7] i.e. things can effect other things through space and time instantaneously (violating Einstein's prohibition of things going faster than light).

Nelson's Stochastic Mechanics attempts to explain QM via quantum fluctuations, in particular, an aether undergoing Brownian Motion [8].

Indeterminate Granular Space-time Mechanics sees Ψ as fluctuations in the curvature of space-time undergoing Brownian Motion. Here, the 'points' of space-time are replaced by 'grains.' In IGSM, in the absence of mass, space-time becomes, not flat as Special Relativity posits, but undefined (stochastic or chaotic). Einstein believed this as well [9].

A problem common to all the interpretations is at what scale does superposition go away? While one might have an electron simultaneously here or there, we most likely cannot have a live and dead cat. IGSM deduces that quantum superposition ceases for masses greater than the Planck mass (about 2×10^{-8} kilograms) while the other models use 'decoherence' [10] ideas to attempt to solve the problem.

What will turn out to be the 'right' (or at any rate, the most compelling) model? Who knows? One tends to hang on to the model one started out with. As Max Planck said, "Physics makes progress one funeral at a time." So we may need to wait for the next generation or two of theoreticians before being able to answer the question.

Now, as to my story: I was working on a physics paper titled, 'Ψ: a Toy-Model, Collapse, and The Schrodinger Brat Paradox.' I'd expanded the Cat Paradox (rather in the way described in the story) in order to argue against the Multi-world Interpretation. (I'd considered MWI to be a 'map is not the territory' issue and a misuse of probability theory.) When I thought of the name 'Brat Paradox,' I thought it would make a good name for a science fiction story. Shortly thereafter, I was invited to submit to this anthology, so I wrote the story. Aside from the central ideas of the story, I tried to keep the science (and comments about some of the scientists) accurate.

Finally, as to the byline: I write my science papers under my actual first name, Carlton, and my fiction under the shortened form, Carl. As I'm not sure if this submission is predominantly science or fiction or both, in the spirit of quantum superposition, I've used both names.

References

1. P. Byrne, *The Many Worlds of Hugh Everett III* (Oxford University Press, Oxford, 2013). Reprint Edition
2. L. Rozema et al., Phys. Rev. Sett. **109**, (2012)
3. D. Bohm, Phys. Rev. **85**, 166–193 (1985)
4. E. Nelson, Derivation of the Schrodinger equation from Newtonian mechanics. Phys. Rev. **150**(4), 66
5. C. Frederick, Stochastic space-time and quantum theory. Phys. Rev. D 13, **12**, 3183 (1976)
6. C. Frederick, Indeterminate Space-Time Quantum Mechanics: A Computer-Augmented Framework Using Wiener-Like Processes, ArXiv:1601.07171 (2016)
7. J.S. Bell, *Speakable and Unspeakable in Quantum Mechanics* (Cambridge University Press, Cambridge, 2004)
8. T. Szabados, An Elementary Introduction to the Wiener Process and Stochastic Integrals, arXiv:1008.1510vl
9. A. Einstein, *Relativity: The Special & General Theory*, 15th edn., p. 155
10. W. Zurek, Decoherence, Einselection, and the quantum arguments of the classical. Rev. Mod. Phys. **75**, 715–775 (2003)

Fixer Upper

Eric Choi

The International Space Station was dead.

From afar, the ISS outwardly looked much as I had remembered it, a long truss structure with four pairs of massive solar arrays and the stacked cluster of dull silver and off-white pressurized modules. But its attitude, its orientation, was wrong. The main truss was tangent to the limb of the Earth, but the gradient of gravity had pulled the stacked modules into a line pointing towards the surface of its planet of origin. As we approached, I saw insulation blankets that had once been pristine and white were now cracked and stained a yellowish brown. The solar panels looked drab and were pockmarked with ragged holes in various places. There were no lights in the windows.

I had been expecting to see this, and in fact the station was in somewhat better shape than I had feared, yet I could not help but be sad. Once upon a time, a lifetime ago it seemed, the ISS had been my home in orbit for five months. To see it like this was heartbreaking.

"*Shénzhōu J-8, kāi shǐ fēi chuán duì jiē,*" intoned Shěnyáng Mission Control over the radio, indicating clearance for approach and docking.

"Shì," replied Commander Yuán Lìxúe. She turned to me. "Kristen, jì xù jiān kòng duì jiē mù biāo fāng xiàng."

As I watched the range and range-rate numbers count down on the control panel, Commander Yuán grasped her hand controller and brought our spacecraft to within two hundred meters of the sprawling, lifeless complex. The rear thrusters fired, and we begin to ease towards the docking port at a snail's pace of a few centimeters per second. At ten meters, a proximity alert flickered on

© The Author 2017

M. Brotherton (ed.), *Science Fiction by Scientists*, Science and Fiction,
DOI 10.1007/978-3-319-41102-6_12

the screen. I overrode the warning, and we continued forward. Lìxúe aligned the white crosshairs on the screen squarely with the alignment target on the docking port.

Our spacecraft contacted the station with the slightest bump. Nothing happened.

I glanced at Lìxúe, expecting her to call down to Shěnyáng for instructions. Instead, she simply pulled us back a few meters from the port, fired the aft thrusters again, and rammed us home harder. This time, mechanical hooks and latches swung into place and locked the vehicles together.

Yuán Lìxúe and I were the first people to visit the International Space Station in over two years.

***

We were docked to the Poisk research module in the former Russian segment of the International Space Station. After going through the post-docking checklist, Lìxúe and I unstrapped ourselves and floated to the hatch at the end of the Shénzhōu's orbital module.

Lìxúe asked Mission Control if they were receiving any telemetry from the station on environmental conditions. They said there was none. This was not surprising considering the station had no power, but it meant we were flying blind. There was no way to know what the environment was like on the other side of the hatch.

We adjusted our launch and entry suits, put on woolen hats, and donned oxygen masks. I threw the switch on my tank, breathed in deeply — and got nothing. The mask just collapsed around my face. I double-checked to ensure the switch was thrown. It was. I sucked in again, harder this time. The mask collapsed further around my face.

"Zěn me le?" Lìxúe asked. Her round open face was wrinkled and her black hair was streaked with grey. Unlike many Chinese her age, she did not dye her hair.

I told her my mask was not working. We went back to the Shénzhōu to look for a spare, eventually finding one in the descent module. I put on the new mask and flung the switch. This time, oxygen flowed.

Returning to the hatch, I grasped a small star-shaped valve and turned it. Holding up a finger, I felt air rushing into the station. The pressure in the Shénzhōu began to fall. Then, the flow stopped. The ISS was still airtight.

We opened the hatch. I dove ahead, but unexpectedly bumped into Lìxúe. We looked at each other. My startled annoyance quickly gave way to embarrassment as I realized my mistake. On this expedition, I was not the commander. I gave way, then followed Lìxúe into the station.

When the air hit my face, I realized how bitterly cold it was. Moisture from my exhalations froze in a tiny cloud around my face. I played my flashlight along the darkened bulkheads of the Russian research module. The beige walls and grey electronics boxes were covered with a thin coating of ice. Mold from past occupations was frozen on the panels.

My gas detector did not register any toxic fumes. I lifted my mask and took a cautious sniff, followed by a deeper breath. The air was very cold, but it seemed to be all right.

"*Zhǐ huī yuan huì bào wēn dù?*" Mission Control wanted to know the temperature.

We looked at our thermometers, and I was surprised to discover that the scale only went to zero degrees Celsius.

Lìxúe suddenly turned, and spat on the wall. I tried to hide my disgust. She looked at her watch, timing how long it took for her saliva to freeze. Twelve seconds. "Líng xià shí dù," she declared. Minus ten degrees Celsius — a rather bracing Minnesota winter, I thought to myself.

We continued into the station, floating through the Poisk module into a small connecting node, where we did a ninety degree turn through a hatchway into the Zvezda service module. I opened the window shades to admit a little sunshine, but it was still terribly cold. With the interior lights out, the narrow fields of illumination from the windows and our flashlights created stark and eerie shadows behind the angular equipment. Sunlight streaming through the small windows lit up myriads of dancing motes and drifting rubbish. A small photograph tumbled by, and I reached out to grab it. The cherubic face of a little boy, perhaps three or four years old, smiled out at me.

Our first priority was to restore power, and with it heat, light, and full life support. Lìxúe and I removed a bulkhead panel and located Zvezda's ancient nickel-cadmium batteries, replacing them with modern solid-state electrolytes. Connecting the new batteries to Zvezda's power bus was a difficult task because we had to take off our gloves, and our hands soon become painfully cold and stiff. The silence was oppressive. With no motors or ventilators whirring, this frozen nook in space was the quietest place I had ever been.

My oxygen mask suddenly collapsed against my face. Startled, I looked down at the tank and saw the display still showing around 40%. I tapped the tank and flipped the switch back and forth, but no more air flowed. Resigned, I took off the mask and continued working. But without ventilators to circulate the air, exhaled carbon dioxide hovered about my face. My head began to ache, my arms and legs grew sluggish, and I started feeling drowsy.

"Nǐ méi shì ba?" Lìxúe said, asking if I was all right. Seeing my mask off, she handed me hers.

I brusquely told her I was fine. It really irritated me, this attitude that Westerners in general and Americans in particular were weak and needed looking after.

The ISS creaked and groaned under thermal stresses as it passed from sunlight into the Earth's shadow. Lìxúe and I retreated back to the relative warmth of the Shénzhōu to ride out the orbital night. Forty-five minutes later, we emerged to resume our work.

The first battery was hooked up, and I saw Lìxúe smile as the voltage rose. The job went quicker for the other seven units. After a final check of the connections, I switched on the main power. Suddenly, the dead module sprang to life. Fans began to whir and the ventilation system kicked in with a low hum. Displays illuminated on several pieces of equipment. The interior lights came on.

"Shěnyáng, diàn lì huī fù chéng gōng!" Lìxúe smiled broadly as she reported our success to Mission Control.

A moment later, everything went dead again.

***

We eventually managed to restore partial main power. This took the better part of a day, by the end of which we were simply exhausted.

But there was one more obligation. Shěnyáng Mission Control told us to standby for a media event in twenty minutes. In my former life at NASA, press conferences were actually something I enjoyed. Outreach and education had been important parts of my job, and I relished sharing the wonders of spaceflight with the general public.

Lìxúe unrolled her tablet and set up a camera and lights on the bulkhead. She said I looked tired and told me to let her do the talking.

I nodded curtly. Because nobody is interested in what the dumb Red Card American has to say, I thought to myself.

"*Huānyíng lái dào CCTV,*" the interviewer began. "*Wǒ shì zhǔ chi rén Dù Tíngfāng. Jīn wǎn, wǒmen… chǎng…jiǎng tàikorén Yuán Lìxúe hé Kristen Bartlett…*"

The rapidly spoken Chinese, exacerbated by the poor quality communications link, was hard for me to follow. Lìxúe spouted platitudes about the importance of our mission, how the station was in great shape (even though we had only been here two days and had not yet ventured into the other modules), and how much she enjoyed working with me (even though the feeling was not necessarily mutual). The interview lasted about fifteen minutes, but it was rough because the comms kept dropping out. Through it all, I stayed quiet with a forced smile plastered on my face. There were no questions for me.

Finally, the interviewer signed off. My fatigue returned with a vengeance, and I was expecting to head back to the Shénzhōu for some much needed sleep when Mission Control called up again. We were told to standby for a message from Liǔ Diānrén, CEO of the Xīn Shìjiè Corporation and the financial backer of our expedition,

Dismayed, I turned to Lìxúe. She didn't appear surprised.

Liǔ Diānrén appeared on the screen. He looked to be in his late thirties with slicked black hair, a pale square face, and an angular beard. Our sponsor was speaking from his office at Xīn Shìjiè's headquarters in Sharjah. I found his Chinese even harder to follow than the CCTV host. He congratulated me and Lìxúe for a job well done, and reiterated his dream of bringing the station back to life and creating a new heavenly dynasty or words to that effect. By the time he started babbling about something called *Cháng'é Bèn Yuè*, he had pretty much lost me. But I caught the last bit, in which he announced that henceforth the module we had successfully reactivated would be known by its new, proper Chinese name — Dōngxīng, the Eastern Star.

Strangely, we didn't lose comms once during the entire rambling monologue. After what seemed an eternity, Liǔ Diānrén finally shut up and the screen went dark.

Within seconds, so did the rest of the "Dōngxīng" module.

***

Dōngxīng was only the start of the name game. As Lìxúe and I reactivated modules, Liǔ Diānrén called up to personally rechristen each with Chinese monikers. The Zarya functional cargo block became Shǔguāng ("New Epoch"), the Unity node was now Níngjìng ("Serenity") and the former U.S. laboratory module Destiny became Wángcháo ("Dynasty"). I found this creeping Sinofication — a cultural appropriation as much as a technological one — quite upsetting, all the more so with the knowledge that I was abetting it.

Power, life support and communications had been more or less stable over the past week, enabling Lìxúe and me to prepare for the arrival of the last three members of our expedition. The lights were dim and the station smelled like a musty old wine cellar, but we tried our best to clean up. We stuffed excess gear and broken-down equipment behind panels, mopped up globs of floating water, scrubbed down the bulkheads with fungicide wipes to remove the mold, and installed new filters in the air cleaning system. It was a far cry from the ISS that I remembered, but things were at least in better shape than what had greeted me and Lìxúe just over a week ago.

From a scarred window in the Dōngxīng service module, I watched the approach of the Shénzhōu J-9 spacecraft to the docking port of the Kēxué laboratory. A momentary shudder reverberated through the station at the moment of contact.

I swam into the Kēxué module. Lìxúe opened the hatch, and the final three members of our expedition emerged.

"Zīchéng! Wénxìn! Chéngfēi!" I called out.

"Kristen!"

The commander drifted aside, apparently allowing me to greet the newcomers first. Fàn Zīchéng was a burly giant of a man with jet-black crew cut hair. It always amazed me that he was able to fit his immense frame into the cramped Shénzhōu. Next was Cài Wénxìn, a graying bespectacled man who in every way appeared the physical opposite of Zīchéng. Last was Zhāng Chéngfēi, a middle-aged man of medium build with a round, puffy face. I had worked with these guys in Shěnyáng for months with the original mission commander, whom Lìxúe had replaced in the final weeks of our training.

They each gave me a hug as they exited the Shénzhōu, before floating over to Lìxúe to shake her hand and exchange a few words. We all then migrated in single file — heads closely following feet — into the Wángcháo laboratory, where we alighted to endure another video message from Liǔ Diānrén.

When this guy talked the comms never failed, much to my disappointment. Liǔ was either speaking slower or the long solitary confinement with Lìxúe had improved my Chinese, because I actually made out most of his rambling monologue.

"*First you were two, and now you are five. I wish you were eight, for that would be of good fortune, but my engineers tell me the station cannot sustain more than five. Perhaps I will fire them! But five it is now, and you must work together as one, like the white shadow that moves unseen across the heavens. You must succeed, for if you do not, the fault will be yours. Work well, you five heavenly shadows!*"

I blinked and shook my head. Perhaps my Chinese hadn't improved as much as I thought, because nothing Liǔ said made sense to me. I glanced over at Zīchéng, Wénxìn and Chéngfēi. The three guys looked excited, occasionally exchanging quiet words and pointing at the screen and nodding. Lìxúe just stared ahead with her arms crossed, her face an expressionless mask.

***

Docked to the aft port of the Dōngxīng service module was Progress MS2-1C, the last in a series of Russian expendable cargo spacecraft that had delivered supplies to space stations since the late 1970s. This particular vehicle had been heavily modified with a larger engine, extra fuel tanks, and the ability to draw residual propellant from the station itself. Its mission had been to deorbit the International Space Station following its abandonment. But in a final act of reckless optimism, the original international partners agreed instead to use the Progress to boost the ISS into a higher orbit, with the hope that a future crew might someday reactivate the station.

It had always been my dream to command such a crew and revive the ISS. Not being in charge was bad enough, but being the minion of a Chinese expedition was something I never would have guessed and don't think I've quite gotten over.

Lìxúe and I were loading broken equipment and garbage into the cargo module of the Progress. It was not a pleasant task. We found ourselves entangled in a chaotic cloud of rubbish, trying to push the floating garbage bags and obsolete electronics into the Progress. I gagged from the smell of decomposing trash.

There had never been much small talk between me and Lìxúe, both because of the language barrier and the fact that I really didn't know her well. But the cold, damp and malodorous Progress was just too much. I needed a distraction.

"What you know of Liǔ Diānrén?" I said at last in my awkward Chinese.

Lìxúe stopped, a garbage bag frozen in mid throw. I feared she might have thought my question impertinent or disrespectful. But then she spoke.

"Liǔ Diānrén lives in a cloud of success. But it's not his success." She tossed the bag into the Progress.

There was an awkward silence. Her answer was curious, and now I really wanted to hear more. "Explain."

Lìxúe hesitated. A strange look crossed her face. Perhaps she thought she'd already said too much, but she continued. "Diānrén's father, Liǔ Fākuàng, had built up the Xīn Shìjiè Corporation from a dumpling stall in Xī'ān into the diversified global corporation it is today. Liǔ Fākuàng was lucky enough to strike gold, but his son wasn't there when he was swinging the pick. Diānrén merely inherited the throne of his father's industrial empire when the elder Liǔ became ill and retired early, a dynastic succession if you will."

I didn't know any of this, and I was fascinated.

"Do you know what is a shǎodì?" Lìxúe asked.

"Little emperor?" I repeated.

"Liǔ Diānrén is like me, an only child — an only son — born of the first generation under the old One-Child Policy. The difference is, my family wasn't rich." Lìxúe paused. "He has a full-time employee whose only job is too peel grapes for him, the way his grandmother used to when he was small."

And I thought my daughter was a princess.

Lìxúe sighed. "We are a nation of badly brought-up children."

The Progress cargo module was nearly filled to the brim. After checking that the seal around the hatch was airtight, we loosened the bolts that secured the Progress to the docking ring of the Dōngxīng module.

"Shěnyáng, we are go for Progress separation," Lìxúe radioed down.

"*Chéng rèn*," Mission Control acknowledged.

Ground controllers commanded latches to open, and springs pushed the Progress away from the station. Floating to one of the small scratched up windows in Dōngxīng, I watched the vehicle back away, plumes of thruster exhaust periodically puffing from the squat insect-like spacecraft. I continued to watch until it resembled nothing more than a distant star in the blackness of space, destined for a fiery reentry over the South Pacific.

Good riddance to bad rubbish.

***

It took weeks to restore sufficient functionality to the Space Station Remote Manipulator System, the seventeen meter long robotic arm that had been crucial to the original construction and maintenance of the ISS. Most of the Canadian engineers who had developed the SSRMS were either retired or dead, but Liǔ Diānrén's team managed to find a handful of the old timers and brought them to Shěnyáng. Despite their best efforts, only one of the arm's redundant control strings would respond to commands, and the wrist roll joint had failed completely. We would also have to make do with the arm's remaining 386-processor, an archaic chip that had been obsolete even when it was originally launched.

"*Confirm* Dàshǒubì *is in the pre-capture position*," Shěnyáng called up, referring to the robotic arm by its new Chinese name.

"Acknowledged." I looked out the Cupola windows, pitted and pock-marked from years of dust and debris impacts. Above the hazy blue of the Earth's limb, the arm hung in space like a giant articulated soda straw, its formerly white thermal blankets now a dull yellow from long exposure to ultraviolet radiation and atomic oxygen. About ten meters below was Huǒniǎo — the Phoenix. Twenty meters long and five meters wide, Huǒniǎo was a cargo vehicle made up of half the payload shroud of a Chángzhēng-5G heavy-lift rocket with the modified upper stage still attached as a propulsion module. Resembling a truncated version of the old Space Shuttle payload bay, it was packed with tons of hardware, equipment and tools we needed to complete the station's reactivation.

"Start capture." My hands tightened over the hand controllers of the robotic workstation. I had operated the robotic arm on my first expedition to the ISS. They say that one never forgets how to ride a bicycle, except in this case the bike had one flat tire and a messed up gear shift. The arm was sluggish and I had to fly a non-optimal trajectory to compensate for the failed wrist roll joint, but eventually I managed to snare the Huǒniǎo's grapple fixture. I then drew the cargo vehicle closer to the station and maneuvered it to the berthing mechanism on the Héxíe ("Harmony") node.

"We have Huǒniǎo," I reported.

"*Gàn dé hao*," Shěnyáng radioed.

"Hěn hǎo!" Lìxúe exclaimed happily. "Well done indeed!"

With the capture complete, I cycled through the robotic arm's cameras trying to find a working unit. One of the boom cameras eventually responded to commands. Through a grimy lens obscured by years of brownish-yellow propellant residue, I surveyed the berthed cargo ship. Most of the payloads were familiar to me: the new control moment gyroscopes, the high-pressure gas tanks for the airlock, the ammonia coolant tanks, the batteries, and…

I frowned. There was cargo I did not recognize from the manifest: folded-up structural members, and oddly shaped tanks of unknown content. I powered down the robotic arm and went in search of Lìxúe to inquire.

The commander was in the Wángcháo laboratory module, hunched over an open avionics rack like a surgeon at work. She looked up from the exposed spaghetti-like mess of wire bundles and electronic components.

"Kristen, good job with the Huǒniǎo," Lìxúe said.

"Xiè xiè." I paused to compose the sentence in my head. "Lìxúe, I see cargo in Huǒniǎo not known from manifest."

Lìxúe looked at me with an odd expression. After a moment, she said, "I will brief the crew after dinner."

***

Meals are of great cultural significance to the Chinese, in which the social interaction is far more important than the eating. I was certainly fine with the latter, but my lack of linguistic fluency was still a real barrier to the former. This was the case during our training on the ground, and nothing had changed in space. So, once again I ate dinner in silence while Zīchéng, Wénxìn and Chéngfēi chatted amongst themselves.

The guys looked terrible. Wénxìn and Zīchéng had been trying to fix a coolant leak in the Dōngxīng service module and must have gotten some fluid on their faces because their eyes appeared the size of golf balls. Chéngfēi had been trying to replace a fan and filter in the Kēxué multipurpose laboratory module and had scratched his arm on something while reaching behind a panel. It looked swollen and infected, a dull shade of blue from wrist to elbow like a botched ink job from a shady tattoo parlor.

As I ate my can of lukewarm chicken sticky rice and tube of watery soya milk drink, I tried to follow their conversation as best I could. Wénxìn seemed to be mocking Yuán Lìxúe's son, who was apparently a thirty-something slacker still living at home with his biological father, fantasizing about becoming a big-shot business person like Liǔ Diānrén but never actually bothering to do anything that might realize such a dream. Chéngfēi suggested maybe the guy should join the PLA, to which Zīchéng replied the only army that would take such a loser was the Americans.

They laughed. I pretended not to understand.

Lìxúe floated into the module, and the guys quickly fell silent.

"Nǐ chīfàn ma?" I said, asking Lìxúe if she had eaten.

Lìxúe shook her head. "No, but thanks for asking. I've been busy." She turned to the three guys. "If you're all finished eating, I need to brief you on a significant change to the mission plan."

Lìxúe unrolled her tablet and mounted it to the bulkhead. "Now that the final hardware elements have been delivered by the Huǒniǎo vehicle, I can tell you the full scope of our mission." She woke up the tablet. "Liǔ Diānrén has a great dream. His ambition goes beyond this station endlessly circling our home planet. He wants to push humanity out to a new horizon again. Liǔ Diānrén wants to send this space station to the Moon."

My jaw dropped. The faces my companions registered similar expressions of surprise.

The tablet showed a simulation of the station's orbit around the Earth. "In three weeks," Lìxúe continued, "a cryogenic upper stage will dock with the station. Following our departure, this upper stage will perform a series of propulsive burns at perigee over the course of a month." On the screen, the station's circular orbit began to stretch out into an ellipse that eventually reached out to lunar orbit. "This will put the station into an Earth-Moon cycling orbit, one that will bring the station into nodal alignment to pass behind the Moon every other orbit, about twice a month. Once established in EMCO, an Advanced Shénzhōu vehicle will be launched to deliver the first lunar mission crew to the station."

We stared in stunned silence. Finally Wénxìn said, apparently for lack of a better question, "This new mission…does it have a name?"

"Wàn Hù," Lìxúe replied.

The three guys looked puzzled, but I recognized the name. Wàn Hù was a Ming Dynasty official. According to legend, he was the world's first tàikōnaut, building a chair with forty-seven rockets in an attempt to launch himself into space.

"Any more questions?" Lìxúe asked.

There was something about the rocket-chair story that bothered me, and at last I remembered. Wàn Hù had blown himself up.

***

With a full crew of five and the tons of hardware brought up by Huǒniǎo, the pace and scale of activities picked up significantly. Over the course of two weeks Wénxìn, Zǐchéng and Chéngfēi paired up in turns to conduct a series of demanding long-duration spacewalks, or EVAs. The first excursions replaced the four failed control moment gyroscopes on the Z1 truss. Once activated, the new CMGs restored full attitude control and enabled the station to overcome the gravity gradient torque that had pulled its axis of modules downward, returning the station to its nominal local-vertical local-horizontal orientation with respect to the Earth's surface. Subsequent EVAs installed structural reinforcements at various locations about the station's exterior: the connection between the Héxié node and the Shènglì (formerly Kibo) laboratory, as well as the connections between the Shǔguāng functional cargo block, the pressurized mating adapter and the Níngjìng node.

I was worried for the guys because even in the best of circumstances EVAs were not easy. Indeed, they could be positively dangerous. During the original construction of the ISS, astronauts and cosmonauts would have trained for months if not years to perform complex assembly tasks like these. We prepared instead by studying written procedures and watching videos of practice sessions conducted in the neutral buoyancy water pool in Shěnyáng. It was hardly ideal, to say the least.

A few days later, it was my turn.

The airlock in the former U.S. segment of the station had been called Quest but was now renamed Xīngmén, which agreeably translated into English as "Stargate." The airlock compartment was worn and scruffy. On a wall was a faded mission patch sticker with some Russian words scrawled underneath.

We put on our helmets and donned our gloves. Lìxúe started the depressurization pump, and the air bled away. I was nervous.

"*Lìxúe and Kristen, you are go for hatch opening,*" said Mission Control.

"Hǎo ma?" Lìxúe asked.

I took a deep breath, and nodded.

Lìxúe vented the residual air before releasing the handle and opening the hatch.

The Sun was rising. Our EVA was planned to start at orbital sunrise so that we would get the longest period of light. The entire planet was spread out beneath us like a giant blue, green and white tapestry.

"Wǒmen zǒu ba," Lìxúe said.

We tethered ourselves to the outside of the airlock and performed a final inspection of our tool and equipment packages as well as our Jīnyì propulsion units. After executing the self-test, the intention light on my Jīnyì went green. Lìxúe's, however, stayed red. She tapped it with a gloved hand until it turned the right color.

I could see faint puffs from Lìxúe backpack as I followed her along the length of the Wángcháo laboratory module to the Huǒniǎo cargo ship berthed at the Héxíe node. Our job was to transfer the conformal water tanks that had been brought up in the Huǒniǎo and install them on the outside of the Níngjìng node. Arriving at the Huǒniǎo, Lìxúe and I unholstered our EVA power tools and set to work. We applied our wrist tethers to eye hooks on the first water tank before releasing the tie-rod bolts with our power tools. Then, we grabbed the tank by handles at each end and pulled the vessel free of its flight support equipment.

"Shěnyáng, Kristen and I have the first tank," Lìxúe reported.

"*Chéng rèn,*" Mission Control acknowledged.

Holding the tank between us, Lìxúe and I flew back along the Wángcháo laboratory to the Níngjìng node, where the Xīngmén airlock was also located. During an earlier EVA, Wénxìn and Zīchéng had removed some of the micro-meteoroid and orbital debris panels, exposing the structural attachment points underneath. Using our power tools, Lìxúe and I bolted down the first tank, which was shaped to fit the cylindrical exterior of the node.

After another seven exhausting hours, Lìxúe and I had installed the remaining water tanks onto the Níngjìng node. We then connected the station's fluid lines to the tanks and also installed an external radiation dosimeter. Our tasks completed, I started making my way back to the Xīngmén airlock.

Lìxúe was not following me. Puzzled, I turned. "Everything good?"

"Yes Kristen, everything's fine," Lìxúe said. "I just have a small task to perform."

Lìxúe pulled a small octagonal object out of a pouch. It was a bāguà mirror. According to Chinese superstition, a bāguà is supposed to deflect bad qì from malevolent outside forces. I thought to myself that, really, there is no qì in space — good or bad — because qì literally means "air."

Lìxúe screwed the bāguà into an unused tie-rod bolt hole on one of the water tanks. She jiggled the bāguà to make sure it was secure, then pulled out two more objects — a decal of a Chinese flag, and one of the Xīn Shìjiè Corporation's crescent moon logo. She stuck both proudly to either side of the bāguà.

"*The most important task of the EVA has been completed,*" said Mission Control, with no obvious intention to be ironic.

There was the sound of a tiny explosion, like a kid's cap gun going off. A spider-vein crack suddenly appeared on my faceplate.

"Shit!" I screamed in English, instinctively putting my hands to my face.

"Kristen!" Lìxúe was with me in seconds, her hands reaching over my suit. She quickly found the control for the secondary reserve oxygen tank and cranked up the flow.

"*What happened?*" Shěnyáng demanded. "*What's going on?*"

"Emergency ingress!" Lìxúe declared. "Kristen's helmet is breached."

"*Get back inside immediately,*" Mission Control ordered. "*Her suit pressure is down to 30.6 kilopascals and falling.*"

"Kristen, listen to me," Lìxúe said calmly. "I need you to take your hands off your faceplate. Please, just for a moment."

Reluctantly, I complied. Lìxúe slapped a strip of silvery-grey material to my faceplate. It took me a moment to recognize the stuff.

Duct tape.

"Put your hands back now," Lìxúe said. As I did so, she grabbed my arm and activated the thrusters on her Jīnyì jetpack to push us the last couple of meters towards the Xīngmén airlock. She opened the hatch, and unceremoniously dumped me inside.

"Are you all right?" Chéngfēi asked over the intercom.

"Yes, we're fine," Lìxúe replied.

When the airlock repressurized, I removed my helmet and took a deep breath. The station had a faintly unpleasant odor, something between gasoline and antifreeze. Lìxúe and I looked at each other. Our sense of smell must have been deadened by the weeks of exposure to the sketchy air.

"Smells like Běijīng," Lìxúe said, and then smiled.

I shook my head and chuckled.

Yuán Lìxúe brought duct tape on a spacewalk. How cool is that?

***

On Chinese New Year's Day, we unberthed the Huǒniǎo cargo ship from the Héxíe node and released it into space. Under command from Shěnyáng Mission Control, Huǒniǎo fired its thrusters and pulled away, heading for a fiery plunge into the South Pacific.

That afternoon, Liǔ Diānrén announced through a press release that the reactivated station was now fully operational. Officially. I supposed it was mostly true. Liǔ declared a holiday for the crew and ordered a celebratory meal, both for New Year's and also as a send-off for the impending departure of Fàn Zīchéng, Cài Wénxìn and Zhāng Chéngfēi.

We set up a dining table in the Níngjìng node, allegedly the perfect location for the New Year's dinner because it was now supposedly shielded by the bāguà against bad qì. Against a bulkhead the color of bad teeth, Chéngfēi and Zīchéng put up a red banner with the characters "mission accomplished."

Much like the rest of the complex, the node was a chilly, dimly lit place. The station was simply underpowered, its solar arrays now providing only a fraction of the power they had generated when new, their photovoltaic cells degraded after decades in orbit. Our installation of newer, more energy efficient electronics and lighting systems offset some of the losses, as did leaving unpowered the modules of the station that we were not be using, but it wasn't really enough.

Once again, our meal could not proceed without the annoyance of one final video message from Liǔ Diānrén.

"First, there was a dream," Liǔ intoned solemnly. *"Now, there is reality. Once, it was called the International Space Station. The name itself was a cruel, humiliating joke. How could the station have been 'international' when China was deliberately excluded, despite the fact that even then we had a space program at least equal to that of the United States?*

"No more humiliation! Today, we Chinese have made another great leap forward into the untainted cradle of the heavens. The guǐlǎo have abandoned the stars. Let them now look up and pay deference to the ultimate dynasty that I have created, and know there is a new order in the heavens. Henceforth, it will no longer be the 'International' Space Station but will be known as Xīn Shìjiè Tiān Zhōu, a new spacecraft for a new world, my gift to humanity, a vessel that will soon go even farther than anyone has dreamed."

Zīchéng, Wénxìn and Chéngfēi smiled broadly and patted each other on the back. Lìxúe and I watched in stony silence. With his slicked black hair, demonic beard and ugly ass yellow-beige suit, I thought Liǔ looked like a James Bond villain. The guy was crazy.

"Gōng xǐ fā cái, my loyal comrades, and enjoy the New Year's feast. Eat lots of food to build up your strength so that you can work even harder next year!"

We dug into our food as soon as the video link closed. On paper, it looked like a feast, an eight-item meal starting with tubes of swallow's nest soup and tins of jellyfish salad, followed by thermo-stabilized duck, fish and dumplings, rehydrated báicài and mushroom noodles, and finishing with mango pudding cups. In practice, everything was lukewarm except the dessert, which was cold and watery. Our food had been like this throughout the mission, but I guess I was expecting better for New Year's.

After dinner, Yuán Lìxúe handed out small red gift bags for each of us. Inside was a cheesy booklet commemorating the Xīn Shìjiè Corporation's 25th anniversary, a chocolate coin in gold foil wrapping, a pair of red socks, and most surprising of all, a mandarin orange.

"Lìxúe, this is unbelievable!" Wénxìn stammered. "Where were you hiding these?"

I put the mandarin to my nose and inhaled deeply. It was not fresh and the smell was faint, but the scent of fruit from the green Earth almost overwhelmed me.

"Xiè xie, Lìxúe," Chéngfēi said with heartfelt sincerity. "I can't think of a better way to end our time together in space."

Three days later, Zīchéng, Wénxìn and Chéngfēi donned their launch and entry suits in preparation for departure. Following established procedure, they would take the older Shénzhōu J-8 spacecraft back to Earth, leaving the newer Shénzhōu J-9 vehicle for me and Lìxúe. The Zhìhuì research module to which Shénzhōu J-8 was docked was cramped for five, so the goodbyes were brief.

"You are hereby relieved of your duties," Lìxúe said.

Wénxìn turned briefly to Chéngfēi and Zīchéng before speaking. "We stand relieved."

Lìxúe and I embraced each of the guys as they passed, exchanging expressions of thanks and goodwill. The three of them floated in turn through the passageway into the Shénzhōu, and the hatch closed behind them.

An hour later, Shénzhōu J-8 disengaged from the Zhìhuì docking port. I watched from the scarred Cupola windows, the rugged peaks of the Peruvian Andes providing a spectacular background to the retreating insect-like spacecraft. I continued to watch until the Shénzhōu was only a point in the distance, hurtling towards a touchdown at the Sìzǐwáng landing site in Inner Mongolia.

***

Our final visitor, the Wàn Hù lunar boost vehicle, arrived a week later. Lìxúe and I monitored its approach from the Cupola. I cycled through the station's external cameras, but only the unit at the end of the P1 truss worked well enough to provide an image. Something was amiss. The camera zoom settings were much lower than I had expected. I looked out a blurry window. My eyes widened as it got closer and its size became apparent. The Sun momentarily passed behind the booster, and we were briefly eclipsed in darkness.

"Huge," I whispered in awe.

"Yes," Lìxúe said.

Now within a few meters of the Dōngxīng docking port, I got a good look at the monstrosity. It was cylindrical, almost thirty meters long and eight meters in diameter, and covered with rusty orange insulating foam. A large radiator panel jutted out from one side. At the nose were rendezvous sensors and a docking mechanism, and at the back was a module with a pair of large engine nozzles surrounding by quads of smaller thrusters.

The Wàn Hù booster ploughed into the back of the Dōngxīng service module. A shudder reverberated through the station. From the Cupola windows, I saw the station's central axis of pressurized modules writhe under torsion like an awakening dragon. Wàn Hù was more than twice the size of any of the station's modules.

"Shěnyáng, we have contact and capture," Lìxúe reported.

We spent the next three days preparing the station for its second, and hopefully much briefer, dormancy. Lìxúe and I closed many of the interior hatches to prevent the whole station from depressurizing in case one of the modules suffered a debris strike. The common cabin air assembly — basically the station's dehumidifier — was cranked up to control moisture and reduce microbial growth. We shut off as many pieces of equipment as we could to lower the risk of fire, but left the water processor assembly active to keep fluids flowing through the lines and prevent stagnation.

The first engine burns to inject the station into the Earth-Moon cycling orbit were to start at the end of the third day, before cryogenic boil-off became a problem. I made my way to the Dōngxīng module and donned my launch and entry suit, then went to meet Lìxúe in the Kēxué laboratory where the Shénzhōu J-9 spacecraft was docked. She was already there, waiting for me, still dressed in a T-shirt, long pants…and a pair of bright red socks.

I understood immediately.

"You not coming," I said.

Lìxúe shook her head.

"Why" I asked.

"I am to stay aboard and monitor the EMCO insertion," Lìxúe replied, "and intervene if anything goes wrong."

"Crazy," I said.

"I'll be fine," Lìxúe said stoically. "Life support is still sized for five people, and there's lots of food and water. Once the station is established in EMCO, the Advanced Shénzhōu will arrive with the lunar mission crew…and I will be here, waiting for them."

"Many journeys through Van Allen radiation belts," I said quietly.

"The transits are relatively brief, and the modified Níngjìng node is an adequate radiation shelter."

"Why you really do this?" I asked.

"My family…my son. We need the money."

I nodded, understanding. I had needed the money too. It really was that banal.

"It's ironic, you know," Lìxúe said. "The name, Wàn Hù. It's not Chinese."

I raised my eyebrows.

"The story never appears in historic Chinese literature," Lìxúe continued. "The first mention was actually in an American magazine, I think in the early 1900s. It's a nice story, but it's not Chinese."

I pondered that for a moment, then said, "I am relieved of duties."

"You are relieved," Lìxúe said.

We stared awkwardly at each over for a moment, then hugged. The embrace was warm and sustained. "Zài jiàn, hǎo péngyou," I whispered.

"Time to go," Lìxúe said.

I saluted Lìxúe, then dove into the hatchway to the Shénzhōu. Inside the orbital module of the cramped spacecraft, I turned. Lìxúe was looking at me. The round hatch began to close, gradually eclipsing her face until it was gone.

***

There was the slightest thump as the holding latches and hooks were released and springs pushed Shénzhōu J-9 away from the docking port.

"Shěnyáng, separation at 11:38," I reported.

The station looked vastly different than what Lìxúe and I had seen when we arrived almost three months ago. Most prominent was the bloated orange bulk of the Wàn Hù docked at the end of the Zvezda service module. The massive Chinese booster was almost two-thirds the length of the station itself, from Zvezda to the Harmony node. Along the main truss, the solar arrays and radiator panels were retracted in preparation for the upcoming propulsive thrust, the largest ever attempted in space.

Alone in the Shénzhōu, I found myself referring once again to the station modules by their original names.

I was now above and behind the ISS. My video survey complete, I fired the thrusters again, ending the fly-around and breaking away from the station's vicinity. In a higher, slower orbit, I watched the massive complex pull away. Squinting at the screen, I thought I saw a small panel fall away from the station.

Half an hour later, the ISS was five kilometers ahead of me, soaring gracefully over the Taiwan Strait.

Over the radio, the commentator in Mission Control was counting down. *"…sān…èr… yī…fā shè!"*

The massive cryogenic engines of the Wàn Hù booster ignited.

"Shùn fēng, Yuán Lìxúe," I whispered. "You're going to the Moon."

"Wàn Hù's two main engines are at maximum thrust," Mission Control reported.

Zooming the Shénzhōu's camera on the accelerating ISS, I saw a pair of fuzzy, incandescent auras streaming from the twin nozzles of the Wàn Hù booster. There was little sense of speed, but this first engine firing was supposed to increase the station's velocity by about two hundred and seventy meters per second.

I took a quick glance at my watch. Seventy seconds had elapsed in the planned three minute burn.

Four seconds later, it all went wrong.

At first, it was difficult to see what was happening. I blinked, not quite believing my eyes that were telling me the central axis of the space station's modules no longer appeared quite straight. The calamity became obvious a fraction of a second later when it visibly distorted into a flattened V-shape, the vertex of the bend near the intersection of the module stack and the transverse truss. With a painful knot forming in the pit of my stomach, I realized that I was witnessing a catastrophic structural failure at the junction between the pressurized mating adapter and the former Unity node.

"Oh, crap!" I exclaimed in English. "Sweet mother of —"

The Wàn Hù booster was still firing as the structural failures cascaded. The line of pressurized modules was now severed behind the old Unity node, like a string of sausages cut by a butcher. Seconds later, the central truss snapped like a twig between the S0 and P1 elements. Then, the former Japanese module Kibo broke away from the Harmony node.

"*There appears to have been a major malfunction,*" intoned the cold voice of the Mission Control commentator. "*Telemetry has been lost.*"

I keyed the radio. "Shěnyáng! Terrible structural failure!" Without waiting for a response, I tried to raise the commander. "Yuán Lìxúe, respond!"

Starved of propellant, Wàn Hù's engines finally went dark.

The space station was shattered into five major pieces, with dozens of smaller bits floating about. Clouds of gas, frozen fluid drops and shards of equipment briefly spewed from ruptured modules until the air was completely dispersed to vacuum. Amongst the debris was a small octagonal mirror — the bāguà, which proved itself completely useless in warding off disaster. In time, the smashed remnants of the station would succumb to orbital decay, bringing down to Earth the mad dreams of a little emperor in Sharjah.

The International Space Station was dead.

Afterword

In the final episode of the 1990s science fiction TV series *Babylon 5*, the titular space station is decommissioned by deliberately overloading its fusion reactors and blowing the place to smithereens. "We can't just leave it here, it would be a menace to navigation," an Earthforce commander tells former president John Sheridan, saying the station had "become sort of redundant" and citing recent budget cutbacks. This is a peculiar action because one would think a massive cloud of debris in the Epsilon Eridani system would be an even greater menace to navigation. A more logical decommissioning would have been to crash the station onto Epsilon 3, the planet about which it had orbited, although I suppose Draal and the Great Machine might have taken offense.

Like its fictional counterpart, the International Space Station (ISS) will also require a suitable retirement at the end of its mission. The current multinational ISS partnership — consisting of the United States, Russia, the European Space Agency, Japan and Canada — have agreed to continue operating the station until 2024, with preliminary discussions underway to potentially extend its life to 2028. Whenever its mission finally ends, the ISS will have to be removed from orbit. This will be the largest human-made object ever brought down to Earth, and doing so safely will be a challenging engineering problem. Nobody wants a repeat of the (mostly) uncontrolled 1979 reentry of Skylab, in which parts of the station ended up hitting a remote region of Australia and resulting in a $400 littering fine to NASA.

One of the cargo vehicles that brings supplies to the ISS is called Progress, a type of spacecraft the Russians have been using since the late 1970s to support their earlier generations of space stations. The current ISS decommissioning plan envisions using a modified Progress equipped with a more powerful engine, extra propellant tanks, and the plumbing needed to use up any remaining fuel in the Russian segment of the ISS. This modified Progress would execute a series of high-thrust burns to lower the ISS orbit and eventually bring the station to a fiery destructive plunge into a remote part of the ocean — likely the South Pacific or the Indian Ocean — as far away from major shipping lanes as possible (see "menace to navigation").

Such was also the fate of the immediate predecessor to the ISS, the much-maligned Russian space station Mir. For a brief time, the Mir actually got a temporary stay of execution. In 2000, a private firm called MirCorp signed an agreement with the Russian space company RSC Energia to lease the station for commercial activities. MirCorp privately funded a 73-day mission to Mir by two cosmonauts later that year, which turned out to be the final expedition. Plans to refurbish the station and raise its orbit were not realized, and the Russians eventually succumbed to political pressure to decommission the Mir and concentrate its resources on the ISS. The fascinating saga of MirCorp is chronicled in Michael Potter's documentary *Orphans of Apollo*.

In "Fixer Upper," the fictional Xīn Shìjiè Company is a kind of latter-day MirCorp. But where would CEO Liǔ Diānrén have gotten a crazy idea like trying to send the ISS to the Moon? Perhaps, somewhere in his misguided youth, Liǔ Diānrén might have read the final report of a student project from the 2003 summer session of the International Space University (ISU). Named *Metztli* after the Aztec lunar goddess, the objective of the project was to define options for robotic and human missions to the Moon using ISS capabilities where advantageous. As part of their work, the students conducted a preliminary feasibility study of sending the ISS and/or components thereof towards lunar space using an Earth-Moon cycling orbit (EMCO). The *Metztli* project is one of the most innovative to come out of an ISU session and its final report is well worth reading (http://tinyurl.com/Metztli-ISU). In 2011, *Aviation Week* reported that some engineers at NASA and Boeing were looking into the feasibility of repurposing ISS modules for potential use at an Earth-Moon Lagrangian point or in lunar orbit.

Both the Americans and the Russians have experience in successfully repairing crippled space stations. Six years before dropping in on the Australian outback, the first U.S. station Skylab suffered significant damage during its launch in 1973. As a result of the mishap a micrometeoroid shield separated from the hull and tore away, taking one of two main solar panels with it and jamming the other panel so that it could not deploy. The first Skylab crew of astronauts Pete Conrad, Joe Kerwin and Paul Weitz succeeded in installing a Sun shield over the damaged hull and releasing the stuck solar array. Equally remarkable was the courage of Russian cosmonauts Vladimir Dzhanibekov and Viktor Savinykh, who in 1985 endured almost two weeks of cold, darkness and physical hardship to bring the Salyut 7 station back to life after a power system failure left it dead in space.

Shortly after the launch of the first elements of the International Space Station in 1998, I remember reading an online post by *Babylon 5* creator J. Michael Straczynski in which he commented on the outward physical resemblance between his fictional station and the Zvezda service module of the ISS. Such linkages between science fiction and our real-life exploration of space are a big part of what appeals most to me about the genre. Historian Roger Launius has called space stations "base camps to the stars," and I cannot help but agree. Now, if only the Centauri would actually sell us jumpgate technology, then we'd really be getting somewhere.

Spreading the Seed

Les Johnson

"We're actually going to the stars."

"About damn time."

"You think so? To me it seems like it didn't take any time at all…"

"That's because you were born after they perfected the de Broglie drive. You're used to traveling around the solar system at nearly light speed. Me? I remember when it took a month just to get to Mars on one of the old fusion ships. Now that was real space travel."

For some reason, Akhil didn't think of his friend as being that much older than himself. He knew better, of course; Raymond only looked young thanks to the rejuve treatments he'd received most of his life. He didn't look a day over thirty even though he was pushing one hundred and twenty. When he really thought about it, he was pretty much an inexperienced brat compared to his friend. He was, after all, only thirty-two himself. And, yes, he had been born after the de Broglie drive had changed, well, everything.

"I guess you're right," said Akhil, looking at his friend as if he were seeing him for the very first time.

"Damn straight I'm right," said Raymond, smiling from ear to ear in his usual, and very disarming, way. This was the smile that frequently landed him his pick of female companions in whichever bar he happened to find himself throughout the solar system.

Akhil only mused on his friend's frequent successes with the ladies for a few moments before they returned to the task at hand — fixing a power junction box that had failed during the full systems testing of the *John Quincy Adams*, the ship that would soon take them and ten thousand other colonists on a

© The Author 2017

M. Brotherton (ed.), *Science Fiction by Scientists*, Science and Fiction,

DOI 10.1007/978-3-319-41102-6_13

one-way voyage to a planet known as *Kepler 186f*, orbiting an inconspicuous red dwarf about 500 light years away. The colonists mostly hailed from North America on Earth and they planned to found a republic on the apparently habitable world that circled their destination star. *Kepler 186f* orbited just 34 million miles away from its red dwarf star, an orbit that put the planet clearly into the Goldilocks Zone, making it not too hot; not too cold.

"Do you think our new home will look as good when we get there as it does in the Planet Finder?" asked Akhil.

"How the hell should I know?"

"You're right, you wouldn't. I just wanted your opinion, I guess. After all, we're going there based on some data taken by the Gravity Lens Telescope. You've seen the data. The atmosphere seems similar to Earth's. It's just hard for me to forget that the light they saw has been traveling through space for five hundred years. A lot can happen in five hundred years. And it'll take another five hundred years for us to get there from here. That's one thousand years! What if there was some sort of primitive culture, pre-industrial, when that light left the planet and now they're in the middle of their industrial revolution? We might be greeted by aliens with technologies as advanced as our own…"

"Do you really believe that alien crap?" asked Raymond, putting the finishing touches on the circuit repair they'd been tasked to perform.

"Of course I do," replied Akhil.

"Of course you do," mocked Raymond, again with a smile.

"Stop it, okay. This is important. We've sent out, what? About thirty colony ships so far? Most have gone places I wouldn't want to go. That's why I didn't apply before now. And the people that are being turned into matter waves and zapped out through deep space won't rematerialize and find out that the world they've been sent to isn't worth shit until they get there and it's too late. What if we arrive and find out the current residents don't want to deal with us primitives?"

"I guess we die."

"Come, on, seriously."

"Well, that's a calculated risk. It's much more likely that we'll get there and find out that the planet isn't as hospitable as we thought. We land, build our settlement and then something happens that causes us to be dead and gone in a hundred years. The universe doesn't care if we live or die; it just is."

"They why are you going?"

"Akhil, I've seen change, lots of change, and I don't like where we're headed. When EarthGov removed the ban on human bioengineering, I thought it was a good thing. We were going to eliminate birth defects, remove cancer genes, and make the next generation of people smarter and more resilient. That happened, along with a bunch of super babies with blonde hair, blue eyes and IQ's at the top of the scale. And people with gills moving underwater to live in ocean colonies on Earth and now Europa. Weird, admittedly, but I could even live with that. But then the mind-numbed laborers started showing up on the corporate mining colonies and the Kuiper Belt survey ships. That's when I knew that human nature hadn't changed and that I'd had enough."

"So you're running away?"

"I guess I am. Our colony will be smaller and we'll be struggling to tame a new world. It'll be a while before we have the infrastructure and the luxury to think about changing who we are and what it means to be human. Shit. I'm starting to sound like a philosopher; I hate philosophers." That infectious grin again crossed Raymond's face.

"We're done here," said Akhil as he put the finishing touches on the now-repaired power relay and began cleaning up the mess surrounding their repair.

Both men tidied up their work area and performed one last systems check to make sure the repair was complete. It was time for lunch.

***

Akhil didn't sleep much that night. He shared a room with his contract-mate, Jamie, and neither was much in the mood to do more than sleep. He and Jamie were in year eight of a ten-year marriage contract that both currently believed they'd renew when the time came. They were very compatible, in personality and sexuality, and neither was apt to go looking for anything better — especially among the ten thousand amidst which they now found themselves after they'd signed aboard the *John Quincy Adams*. Akhil's thoughts included a musing on the meaning of a ten-year marriage contract that would expire while their atoms were converted into matter waves and sent across the universe on a 500-year voyage. For them, their last two years under contract would be spent on a new world while back on Earth the legal documents they sealed with their retinal ID's likely wouldn't even still be accessible. After all, the technology to create and read them would be half a millennium out of date. The ship was set to depart in less than a week, and he was having second thoughts.

As he listened to the rhythmic deep breathing of his wife beside him, he thought about the situation that brought them to this place and time and about what their life would be like just next week, their time, when the ship's onboard de Broglie drive dropped them from wave-space and back to particle-space near their new home. He knew enough about the flight plan to know that they'd drop into particle-space a few Astronomical Units from their destination and rely on the ship's fusion drive to cover the remaining distance. A 'conversion safety factor' they called it. He guessed there was some uncertainty in navigating across the galaxy. Hell, yes, he was nervous. And he *was* having second thoughts. Would he and Jamie be here if things had been different? If their daughter Chandana hadn't fallen victim to the latest designer virus cooked up by the disaffected separatists on Luna? If only they hadn't taken that trip. If only they'd taken the trip two weeks sooner. If only… Akhil nodded off and dreamt. They weren't restful dreams.

***

"Listen up people!" Captain Lajoie's voice boomed into their cochlear implants to get the assembled crew's attention away from the hundreds of whisperings and small conversations that were happening in the auditorium that contained nearly everyone who'd signed on to the *John Quincy Adams*. The auditorium was dockside on Neried, which has become the solar system's port of departure for Kuiper Belt exploration and the growing number of interstellar arks. No one was quite sure why Nereid had become the waypoint, but those who came through never failed to be awed by the view of Neptune afforded by the shipyards there. That giant blue ball, taking up an impossible amount of the visible sky above everyone's heads was simply too awe inspiring to ignore. At least at first, until the inevitable call to work sounded and the day-to-day tasks that make up a person's life had to be performed.

Akhil and Jamie had been talking, of course, and quickly stopped to hear what Captain Lajoie had to say. Jamie, in that way that first attracted Akhil to her, raised the left side of her lips into a half smile and winked at him. A sign between them that all was well. Before Akhil completely turned away from his wife, he again appreciated her beauty and gave a silent thanks for her to whatever deity might be eavesdropping on his thoughts. To Akhil, she was the most beautiful woman in the room.

"This is our final all-hands briefing before we board the ship and say goodbye to the Terra System. There's a lot to cover, most of it relates to the final ship checkout before departure, but it shouldn't take more than an hour or so. We've been through most of it before." Lajoie, who didn't appear to be a day over fifty, probably thanks to rejuve treatments, sounded experienced and sure of himself. Akhil was reassured that he sounded competent, but not cocky.

Lajoie continued, "But before we have that briefing, I'd like to introduce Dr. Sakiko Murata. Dr. Murata is responsible for EarthGov's interstellar colonization effort and she has some news to share with us. Some exciting news." Lajoie bowed ever so slightly as he deferred to Dr. Murata who was walking toward him on the stage. Murata bowed in return as she moved front and center on the stage. She looked as youthful and ageless as Asian women always appeared to Akhil's untrained eyes.

"Thank you Captain Lajoie for allowing me to speak to your crew and passengers before you begin your historic voyage. Until your departure, no human ship has attempted to explore or settle more than a few hundred light years from Sol. You will be traveling five hundred light years to what appears to be a destination mostly-habitable for human life. Only you will know the real circumstances there because by the time your initial reports return, thanks to the speed of light limits, one thousand Earth years will have passed. I will be honest with you. No one knows what this solar system will be like after that much time has passed. But you know this already or you wouldn't have been allowed to sign up. I know about the intense screening you all went through to be here today. I know you are ready for both the trip and for taming a new world. We have received some information that you need to know before you depart. Information that some at EarthGov debated, at the highest levels, whether or not to share with you."

Akhil heard several thousand people shift in their seats and more than a few murmur in surprise. He, too, shifted as he stole a glance at Jamie to see her reaction. She was glancing at him. Yes, they had been together enough to be essentially of one mind and each knew how the other would react in most any situation. By unspoken agreement, they both redirected their attention back toward the front of the auditorium and to what Dr. Murata might say next.

"As you may recall, though we have only been sending colony ships like yours to promising destinations for about twenty-five years, we have been sending de Broglie drive-powered scout ships out for nearly fifty years. That means we are now getting direct communication back from some of those scouts who've explored other stellar systems out to about twenty-five light years from Terra. As expected, most of them have encountered scientifically interesting worlds, but few that are even remotely suitable for human life. Thankfully, the only colony ships we've sent out have been to planets deemed spectroscopically suitable by the Gravity Lens Telescope and the other Terrestrial Planet Finder systems. But most of them lie beyond, at distances too vast for either the colonists to have arrived yet, or for those closer in, for their first radio reports to have made it back here." She paused and reached out to one of her aides, who handed her what looked like a small cup of tea. After an inaudible sip, she again looked at those assembled.

"I would like to share news from one of the early survey ships. The news is… disconcerting. If you would set your retinal implants to channel C, I'm sending an image taken on *Gliese 667Cc*, a mesoplanet with an Earth Similarity Rating of over eighty percent that is just less than twenty-two light years away. This rating is only slightly lower than your destination, making it an early target for sending a survey ship. I am sure you have heard in the news the results from the survey, after all, it was one of the first exoplanets with confirmed extraterrestrial life and the journals have been having a field day with the data coming back from the survey ship which is now on its way home with physical samples. The last of the radio data will arrive only a few weeks ahead of the ship as it comes of out de Broglie space and returns home." Murata again paused for a sip from the cup, which she then handed back to her aide.

"What we did not release to the public, at least not yet, are these images."

Appearing on Akhil's retinal implant was a three dimensional image of what appeared to be a city, or the remains of it. It looked a bit like what remained of Abu Dhabi after the city was destroyed in the final Middle Eastern Conflict of 2075. The tactical nuke had devastated the city's ultramodern skyscrapers, leaving them a mangled heap of steel and fused glass lying in the desert. Little or no vegetation had regrown over the city in the intervening decades, as had occurred in the equally devastated Beirut. But Beirut wasn't in the desert like Abu Dhabi, and vegetation had invaded the destroyed city, slowly reclaiming it from the destruction wrought by humans. It was impossible to tell anything about the nature of the previous inhabitants of the alien city from looking at the mangled towers and destruction, but was obvious that it had been created by intelligent beings and then destroyed, likely violently, by those selfsame or other intelligent beings.

"Isotopic analysis confirmed that the city was destroyed by a fusion weapon, probably well over a thousand Earth years ago. The planet is covered with such remains. We have been extremely careful to keep this news out of the public reports until now."

Akhil's retinal implant was now barraged with image after image of devastation — one destroyed city after another. Some were overgrown with plant matter, others, like the first city, were not. He was on the verge of sensory overload when the images stopped and his attention was redirected toward the speaker.

"The survey ship found similar bombed out remains on the planet's only moon and on the moons of some of the gas giants elsewhere in the system. The race that built these cities either committed suicide on a massive scale or someone, or perhaps something else committed genocide."

The room erupted into multiple, competing conversations and the news began to sink in. Dr. Murata appeared to be content to allow the moment of shock to continue among those assembled; she simply stood and stared, watching the reactions of the people in the room. Akhil briefly looked at Jamie and then motioned for her to join him in watching Dr. Murata on the stage. He could tell that Murata was expecting this reaction and had planned her talk to include it. He was beginning to dislike Dr. Murata.

"May I again have your attention? Good. Thank you. We decided to share this information with you so you would know that one of the greatest scientific questions in the history of humanity has finally been answered: we are not the only intelligent, tool-using species to have existed in the universe. We have been searching for centuries for signs of life in the cosmos using virtually every type of astronomy you can imagine only to hear or see nothing. Until now. Now we know that we are not unique. There were others. Others that undoubtedly pondered their place in the universe and wondered about the nature of their own existence. But something happened to them, not all that long ago, and now they are gone. Based on the data from the survey ship, they were a technologically advanced civilization that were close to becoming masters of their own solar system when catastrophe befell them, and now they are no more. From what we can tell, they had not yet discovered a means to become an interstellar civilization and whatever caused them to be destroyed, destroyed them completely as a species. Which brings me to you and why I am here today sharing this information with you as you are about to begin your journey."

The audience was again completely quiet; Akhil and Jamie, like the others around them, sat in rapt attention.

"The residents of *Gliese 667Cc* are extinct because they did not do what we, you, are doing. Spreading the seed beyond the solar system that gave them birth. You are the survival of our species. By taking humanity and Earth life to other stellar systems, that which is humanity shall not go extinct if something terrible befalls Earth or our home solar system. You and the ships before and after you will assure that. Thank you and Godspeed."

With that, Dr. Murata didn't allow for any questions, she simply strode off stage and out of sight.

Captain Lajoie walked back to the center of the podium, took a sip of water from the container he had in his left hand and began speaking, "What you just heard will become public after we launch. Dr. Murata said that EarthGov plans to release publically the news about the remains found on *Gliese 667Cc* next week, the day after we depart. They felt they should inform us before we go and, I for one, am glad they did. What we're doing here isn't just motivated by politics, which is why many of you signed on, and believe me, I know — I've read many of the psych reports." Lajoie paused and smiled as the expected guffaws ripped through the audience and released some of the tension.

"Now let's get on with our final departure briefing before the news of about how important you all are goes to your heads and you can't make sure your jobs are done correctly…"

Akhil, like many others in the auditorium, were already completely distracted by the news from Dr. Murata and had a great deal of trouble paying attention to Lajoie. But Lajoie probably knew that already.

***

The departure of the *John Quincy Adams* was anything but spectacular to those watching from the remote cameras at the departure point. Once the mighty ship had left dock on Neried, it pulsed its fusion drive to reach the departure point a mere 100,000 miles away where it performed its final systems checks. Shortly thereafter, Captain Lajoie turned on the de Broglie generators that converted his mighty ship to a matter wave and sent it on its way toward its new home many light years away. There was no flash or distortion of spacetime that was so popular in the entertainment feeds; instead, the ship was simply there one moment and gone the next. Given the energies required to make the conversion, even those who built the de Broglie generators were often surprised and sometimes disappointed at how unspectacular the process was. To those who rode the ships and whose lives depended on their functionality, boring was good.

Dr. Murata was with her staff on Neried when the *John Quincy Adams* departed. Her aide, the woman who had brought her tea laced with a mild mood enhancer while she was speaking to the crew and colonists just four days previously, was again at her side.

"Sakiko, don't you think you should have told them the rest of the story?" asked her aide, head slightly bowed as if she were ashamed, or hesitant, to question her boss's actions.

"Yes. But telling them the rest of the story would make the news leak we are battling even worse. I'm sure hundreds of them shared my revelation about *Gliese 667Cc* with their families and friends, making the official news release on Monday even less shocking than it would have otherwise been. But that part of the story had already leaked or we wouldn't have told them that much. How could we tell them that we have found similar ruins on two other worlds only slightly farther out? Two other civilizations, neither of which as advanced as the one on Gliese, all bombed out of existence? If we had told them what they might be headed into, they might have mutinied and not gone."

"And by not going, they might contribute to the eventual extinction of the human race."

"Exactly. Whatever unleashed the devastation on our stellar neighbors might still be out there. While we prepare Terra and the solar system for what might very well be a fight for our own survival against whatever is out there decimating those other worlds, we need to get our seeds spread as far and wide as we possibly can."

"But what if they encounter it, whatever *it* is, at their destination?"

"What if they do? Would telling them a few days or weeks before departure make any difference? These first generation colony ships are not battleships. They are designed and built for colonizing. We will not have any armed ships ready for at least another five to ten years. It takes time to retool and become a wartime economy — especially when we haven't even yet made public that we need to become one. No, Kiko, it's best to let them have the delusion that they are only going to have to fight nature out there — not a war. Maybe they will get lucky and find that their new home is both hospitable and out of harm's way."

"Maybe, but then again, we'll never know."

"No, we will not. And when they arrive and listen for news from home, let us just hope that someone is still here to send them some."

Afterword

I owe my career at NASA to the sense of wonder instilled in me by watching Neil Armstrong walk on the Moon and, yes, by *Star Trek* (in all its incarnations). And I am not alone; an informal survey of NASA employees about ten years ago found that a significant fraction were inspired to study science and engineering by Gene Roddenberry's creations. And, if the technological landscape around us is any indication, the show inspired many scientists and technologists in other fields as well. Who didn't pretend that the original flip phones were communicators? Why did those 3.5 inch floppy drives look so much like the computer memory cards used by Mr. Spock? Aren't all those 'health' apps on your tablet really just trying to turn it into a tricorder? And I am convinced that 3D printers are really first generation replicators — "Earl Grey, hot" anyone?

All this brings me to the first of three science topics I incorporated into *Spreading the Seed* — The search for exoplanets. Before ~1992, the only people who knew there were planets circling other stars were readers and watchers of science fiction. It was in that year that the first scientific evidence of a planet circling something other than our sun was accepted as fact. *PSR B1257+12* is a pulsar located 2300 light years from the sun and around it orbit three planets. (A pulsar, or pulsating radio star, is a rapidly rotating neutron star that emits somewhat focused electromagnetic radiation.) Today, there are now more than two thousand confirmed exoplanets — with more being found and confirmed each year.

The second scientific topic in the story is the spacecraft propulsion system used by the *John Quincy Adams*: the de Broglie drive. Like many physics students, I was initially fascinated by quantum theory and its philosophical implications. This fascination has gradually transformed into a reluctant acquiescence to the belief that our macroscopic brains are not well equipped to help us understand the rules that govern the physics of the very small — except through mathematics — and that our common sense may not always apply. And then I was introduced to Louis de Broglie and matter waves.

de Broglie conceptualized the idea that matter, like light, has wave-like properties. This has been confirmed experimentally with electrons, protons and even atoms. The wavelength of a particular bit of matter is inversely proportional to its momentum (mass x velocity). Therefore, the de Broglie wavelength of a desk, human or a spaceship is inversely proportional to its mass — which makes it very, very small. From a practical point of view, this means that Quantum Mechanics, usually applied to individual atoms or groups of atoms, which, under specific circumstances, allows a particle's wave function to relocalize to a different point in space in a process called tunneling, also can apply to large collections of matter — again, like a person or a spaceship. The wave function is just extremely small and the probability of all the atoms making up the person or spacecraft relocalizing someplace else is so small, under normal conditions, as to be essentially zero. BUT, if we can find a way to forcefully relocalize a macroscopic object's wave function, then we would have a working matter transmitter — without the usual wormhole trope. I chose to not reinvent the *Star Trek* transporter where some external device transforms a material object into information which is then beamed and recreated. Instead, I postulate that someone will figure out how to build a shipboard device that transforms itself and anything connected to it into a matter wave that can then be sent on its merry way across the galaxy — limited, of course, by the speed of light.

The last science topic implicit to the story is the Search for Extraterrestrial Intelligence (SETI) to which I have coupled an element of philosophy — the Fermi Paradox. It has been assumed for nearly a century that the natural laws that produced the Earth and life upon it are universal. There should, therefore, be other tool-using, sentient life elsewhere in the universe. Furthermore, given the age of the universe, some of this life should surely have built the capability to explore, colonize or make its presence known across interstellar distances. After all, given our rate of technological progress, many believe we will have a similar capability in just a few centuries. This is the philosophical basis upon which active searches for extraterrestrial life have been based. To-date without success.

This lack of success brings us to the Fermi Paradox. Attributed to the physicist Dr. Enrico Fermi, the paradox which bears his name asked the inevitable question that arises when one ascribes to the philosophical basis of SETI. If intelligent, tool-using life is not unique to Earth and should, therefore, be common throughout the universe, then why don't we see any evidence that they exist? I've lost many a night's sleep pondering this question, as have many of my friends and colleagues. *Why not indeed?* Are the processes which produced and allow life to flourish on Earth more rare than we thought? Is intelligent life self-limiting through war, environmental disaster, or natural catastrophe? We simply don't know. Being a science fiction fan and unrepentant *Star Trek* universe 'believer,' I choose to think we're not alone and just haven't yet found our celestial neighbors — whether they are friendly or unfriendly has yet to be determined.

The Gatherer of Sorrows

J.M. Sidorova

"Pick up and go," the security detail says. The Detail. That's what she calls the man, inwardly. To his face she calls him by his first name. "Hello Jake," she says, "How are you today?" while she finishes grinding petunia flowers in a mortar and then splashes some vodka into the grounds. The petunias, she's been growing on windowsills in coffee cans and in hemp baskets lined with trash bags. The vodka, she's brought from home in a stainless steel hip flask.

A man of a few words, The Detail nods toward the door. Her students, a dozen die-hard ten-year-olds, are watching him, their faces consolidating into a collective *I didn't do it and have no remorse if I did* expression.

"I am in the middle of a lesson," she says. "We are making a pH indicator." With a cocktail straw, sealing its top end with her index finger, she picks up some of the petunia extract and dabs it out onto a paper towel making rows of dots, three dots for each kid in her class.

"Now we let it dry for a minute and then it's ready," she says. "Ladies and gentlemen, have you brought your water samples?"

The kids stir (they like those moments of *Ladies and Gentlemen*, their one and only theater of conspiratorial discovery); some of them open their bags and begin pulling out small canning jars, plastic bottles, tubs. Here she planned to give them a little spiel about controls and tests, and how these matter if one is to do the science right, how, better yet, tests need to be blinded to avoid anticipation bias… But the truth is, the kids are ill at ease: hands

M. Brotherton (ed.), *Science Fiction by Scientists*, Science and Fiction,
DOI 10.1007/978-3-319-41102-6_14

tentative, eyes straying. It's The Detail. The two-way com on his jaw and chin, like a cross between a tattoo and a circuit board from the old days, is enough to make anyone wary, never mind his video intake. He hovers not a foot away from her chair. He's never before showed to pick her up from school, in public. It's always been in the afterhours, in the cavities of her private life.

"Jake, you need to leave the classroom and wait till the top of the hour," she insists. "What is the urgency?" She is tempted to say out loud, *Can't Rollie wait for an extra forty minutes? Is he THAT desperate?* She is tempted precisely because little Cory, always the troublemaker, blabbers, to sparse snickers, "What did you do, Miss Shelley, kill someone?" And more so because Ela-Jo, emotional IQ through the roof, announces, "Today I don't want to learn about our water, Miss Shelley."

Not in front of the kids. Their old Miss Shelley will stick to her pact of silence. She wants to continue coming back here, to this shabby school.

The Detail nods toward the door.

"Miss Shelley, you better go," Ela-Jo advises. "I can run ask for a substitute."

The rest of the kids are silent now, waiting unhappily. And so their Miss Shelley, who wasn't always aged seventy-four, who used to be called Leni, and then Lenora, and then Dr. Mireles, and then the Gatherer of Sorrows, picks up her paper towel with rows of purplish spots, three by twelve and perfectly dry now; she folds it twice and stuffs it in her pant pocket, and then finally, goaded by Jake The Detail, she goes to an empty lot before the school where a sleek black-and-silver pteroglider is waiting to fly her to Rollie.

The machine lifts and banks right; the school's tar rooftop falls away, then rundown townhomes covered in sponsors' ads, then the subdivision, the green separation zone, the corporate park. She settles for a two-hour ride.

***

In Rollie's mansion, an elevator is the only way of getting around if you are let in through the guest entrance. The elevator is operated by four square buttons which themselves make a square. They are marked by colors only: red, yellow, green, and blue. The happy, elemental colors. Jake presses the blue button. There is a hint of motion, then a suggestion of a pause. Rollie's voice through a speaker says, "Hold on a sec." The elevator doubles as a waiting room.

They hold on, are held in. Finally, the door slides aside and there is Rollie, wearing a French terry pajama jumpsuit zipped half-way and folded back, his wet, naked top half is steaming, his hands are each holding a long flute glass, full and fizzling. "Lenora!"

"Roland," she says, stepping forward.

The hall around her resembles a Turkish bathhouse — mosaic tile, water reflections dancing on the ceiling. Water is everywhere, dripping, condensing, trickling. So much water. There is a bar (wet, naturally), and nearby a vast curving couch with water-repellent cushions, water beading copiously on all of them. Some half-wet towels are heaped on the floor between the couch and a large pool of shimmering water. "Take a seat," Rollie says. He is not exactly smiling, in lieu of a smile he widens his eyes in an overexcited, hyperthyroidic kind of way. He is in fine health, of course, a superb thirty six-year-old exemplar of a natural leader. And he is Rollie. His facial expressions can mean anything.

Lenora sweeps beads of splashed water off a cushion. She can't help but rub together her moistened fingers. The water is not soapy, not tacky, leaves no residue. Rollie watches her with transitory curiosity, sipping out of his flute. "It's good water," Lenora says. "Very nice water."

Rollie shrugs.

"Why here, why not in the… upstairs, I guess?"

Another shrug. "Just because. Don't know. Sit. No, *sit*. Tell me how you met my father and uncle."

It's going to be a long day. She sits down, feeling that leftover droplets are seeping in through her pants. She sucks in a mouthful out of the flute — a mojito, of sorts. She holds it till it tingles, then swallows. "I've told you a zillion times."

"Tell me again."

"Can I have something to eat with this?"

"Not yet."

"I met them at a conference," she says, while memories of opening her story with these same words on so many previous occasions amplify like in an infinity mirror, with Rollie and her getting progressively smaller. Younger. Down at the bottom of that mirror Rollie is a little boy, all tucked in for the night, all eyes and ears, and her story is a bedtime fairytale. So long ago. Now the little boy is all grown up and the fairytale has become a strange, strange ritual.

***

She met Yric and Paul Benes almost fifty years ago at a conference organized by their Life Science Foundation for the recipients of the Foundation's grants. She was just twenty-eight and already a junior faculty at Stanford. She knew nothing about anything except what she wanted — a single, simple thing: her science was the best, the coolest and worthiest, and she was ready to bust her ass for it; all she asked was that the world *just let her do it*. That's all.

Yet by the time of that fateful conference, she had been locked in a three-year-long siege of the National Institutes of Health over funding, lobbing her grant applications at them, watching the applications bounce, rejected. Three years and all she had to show was a small grant from the Benes Foundation. She *had* to do better than this.

She'd read how-to guides. Miracles did happen, said the guides, instances of capturing an ear of a philanthropic billionaire who itched to finance some bleeding edge bioscience *were* known to occur in the real world — not just in the movies. Even the top tier journal *Nature* said so. She'd read that article: polish your elevator pitch, *Nature* said, hone it down to an overawing science one-two punch-out delivered in under the five seconds it takes to zoom from one floor to the next in a hypothetical high-rise teeming with billionaires. Practice it on anyone who cares to listen and on those who don't.

She practiced on cashiers at grocery stores.

***

Seated on a wet cushion in Rollie's bathhouse, holding her mojito on an empty stomach, Lenora the seventy four-year-old stares into her infinity-mirror memory. "I came to the conference to seek them out," she says. "I found them and made my pitch."

***

She'd bought what she thought was a sharp business suit. She'd looked up Yric and Paul's headshots (boyish-looking identical twins, smiling open-mindedly) and had extracted out of one of the organizers a priceless admission that they would be present at the conference. She kept her eyes peeled at all times. Even while she was giving her talk. It was a good talk; she'd rehearsed it obsessively. It was witty, and inspired, and confident. But she did not see Yric or Paul, only the Foundation's management was present. After the talk she hid in her room with a searing headache, popping Tylenols; then kicked and shoved herself back out the door to attend the closing banquet. In the lobby a man in jeans and a hoodie approached her. Only up close she realized it was one of them, the twins. "Dr. Mireles? I am Yric Benes. My brother and I are very interested in your research. Would you be willing to tell us more about it?"

She *was* willing, and it entailed getting in an elevator with Yric right there and then, and riding all the way up. The conference was in a hotel. The hotel had a penthouse. Penthouses were billionaires' natural habitat. Almost too hackneyed to be true.

***

Rollie chimes in: "You thought it was creepy, remember?"

Creepy, as in going with an unfamiliar man to his hotel suite. Lenora frowns. Once, years ago, she confessed this to Rollie. She didn't know that Rollie would latch on to it, peel it of all context. He likes to taunt her.

"No, I didn't, why would I?" Lenora says.

***

Yes, she did. A background hum of apprehension not untypical for a young woman. She was used to it. It wasn't a big deal. It shared stage with every other thought, including: she was thrilled and flattered. And at the same time she thought how pathetic and inadequate she looked, or how Yric seemed like *her* kind of guy, a nerd, if a bit older and more baggy-eyed than his headshot claimed. And all the while she heard herself talk normally — not talk up, that is, she even chortled in her usual, slightly off-kilter manner and in all the wrong places. Maybe because she didn't even believe that all this was actually happening. Yric said they were never in the audience, that they watched the proceedings over a closed circuit TV. "Oh, I see. Of course," she said goofily.

"No, it's just that you science types are so… formidable," Yric said. "We just didn't want to look like idiots out there."

Formidable? Idiots? Said by one of the two geniuses who had hacked together a search engine algorithm that had… pretty much changed the world! Leni just about buckled her knees, like a fan-girl.

And there it was, the penthouse. Paul, another nerd in a hoodie, rose to greet her and said, after a handshake, "So, intergenerational epigenetics, huh? The sins of the fathers, visited upon the —"

***

"— sons?" Rollie knows it by heart. He always asks for more details but he never understands them right. How can she make him see that scene the way she'd seen it: a big-screen TV with a still of Leni pointing at a slide, frozen in motion, and in front of the screen two identical guys in identical T-shirts that said *Back off, I'm doing science!* which they purposely wore, they explained, for their closed-circuit TV "science seshh"…

Rollie is sprawled out on the couch opposite Lenora, idly tying, untying a knot of his jumpsuit's sleeves over his belly button. The side of Rollie she gets to see is nothing like Roland Benes the Federal Secretary for Information and the man who controls one of the biggest fortunes worldwide — the face the rest of the world is used to seeing. And that is what makes Lenora so ill at ease.

How different he is from his… parents. "Yes, that's what your father had said."

Rollie stops fumbling with the sleeves and gives her a stare. *"Do sins of the fathers visit upon the sons? Or is the slate wiped clean with every generation?"*

He is quoting from her talk, as Paul was quoting then. She chooses to react to it as if it weren't a question. She smiles. "I was young and cocky."

***

"The Lamarckian adaptive inheritance," Paul continued, "Oooh, the heresy!" He flicked his hands in a half curly-quotes, half voodoo gesture. Leni couldn't yet understand if he joked or challenged her. Paul concluded, "Good stuff. So thick, you can stir it with a stick, right?"

"Um, sort of," she said. "But when people say something is a heresy it implies there is a dogma. We don't really have dogmas."

They smiled like one. "We don't either."

She smiled back. Her headache was receding. "We just have things that we understand better, or less well, or not at all. And things that we suspect but can't yet prove, because we don't have either the tools or the conceptual groundwork. Like way back when, everybody could observe electrolysis in water but no one could explain it for almost a century. This had to wait till they started thinking in electrons and protons. Same here. We all see that the information about one's important or enduring life experiences gets transferred to one's progeny. Systemic starvation or even psychological stress in fathers does result in altered physiologic and behavioral traits of children — this has already been observed. It makes sense too: if our environment triggers a persistent condition, we better respond, learn, adapt, and pass the lesson down the line, right? The point is we just don't know yet how it happens, exactly. What the mechanism is."

"But you have the tools," Paul said.

"Getting there."

"And the groundwork," said Yric.

"We do now."

"And it is?"

"The discovery of small noncoding RNAs that latch on to our DNA." The brothers looked like she needed to explain. "Noncoding means they don't get translated into proteins. They have other important functions. It's like if the DNA is a letter of the law, like a constitution, these RNAs are like the practice of it, they are suggestions and interpretations, they are memos on how to apply the law on a case-by-case basis. More — or less — stringently. Making exceptions — or not. Sorry," she said, responding to their faces, "seems like the research we do just begs for all these… loaded metaphors. I mean, they are just that. I am popularizing." None of this qualified as a good pitch, she thought with alarm.

They laughed. "No, it's cool," Yric said. "We get the difference. Go on. What is *your* groundwork?"

Her cheeks felt hot and her heart raced. This was it, make it or break it. "I… well, to continue with the metaphor, imagine a law that is interpreted and practiced a certain way for a while. Then one day you go look up the actual letter of the law — and you find that it had changed in accordance with the practice. The act of practicing it changed it in the book, you see? That's my groundwork. I know how to look for the RNAs that are the practice of the law, and I have caught a few in the act of tweaking the actual law on the books — in the DNA. But that is *your* RNAs acting on *your* DNA, not on the DNA of your future next generation, the DNA that is sequestered in your germline. Now. If I am able to continue my work, I will provide a gold standard proof of intergenerational inheritance. This is how. Step one: I will induce a specific RNA into existence. Step two: I will satisfy the necessary and sufficient criterion: that I can isolate this RNA and transfer it from one individual to another. Step three: I will demonstrate that this RNA tweaks the DNA in the germline and in the next generation."

***

Abruptly, Rollie springs up and heads for the pool. Lenora finishes telling her story to his bare back, "I explained my research to Yric and Paul. They played back my talk, which they recorded, and we went over it in detail. They had good questions, and were so fast at grasping things. Terrifically smart. And we talked like equals. They were always very… egalitarian this way."

"Were they," Rollie mutters. "And then they gave you the money."

"Not right then. When we parted they thanked me for explaining the science so they finally understood it. A letter came in the mail a month or so later." She has never told Rollie about her "the letter and the practice of the law" metaphor, and she thanks her lucky stars for that omission. For quite a while now, that metaphor has seemed ominous to her.

Standing on the edge of the pool, Rollie drops his French terry jumpsuit, steps out of it, buck naked, and dives in. Splatters hit Lenora on the face. She sighs. "Roland, what's troubling you? Can I help?" She hopes the words do not come out as rehearsed as she feels they are.

"Not now," he responds.

"How is your father doing?" This is genuine.

Rollie is entering a program into the pool's panel. "Daddy Paul?" he remarks between key punches. "His nurse reports he is saying he is Yric."

As if on cue, Lenora's stomach feels like it clenches into a fist and pounds the bottom of her diaphragm, then recoils, aching. She suppresses a cringe: Rollie glances her way, appraising. *That's it*, she thinks. That is why Rollie has summoned her. Another poke under the diaphragm. Then: *could it be?* Then: how hopeless, ridiculous, pathetic of her to even consider this! She *knows* it can't! All the while, she does her best to appear unaffected. She mustn't be dismissive but she can subvert, undermine. "Well, in his state such claims should be taken with a grain of salt. Don't you think?" She wants Rollie to reply but he doesn't. This means he will revisit it on his own terms.

"Daddy Paul" Benes is eighty-three and has Alzheimer's. She hasn't seen him in years. *A billionaire recluse* to all, but Rollie may well be keeping him locked away somewhere: image control. Rollie slams a button on a pool's panel; the water in the pool rolls into motion, Rollie swims against the stiff current, staying in one place. He shouts over burbling water, "Which one did you like better, Yric or Paul?"

Yric, always Yric, she tells herself while pretending she did not hear the question. It works. Swimming distracts Rollie. "Don't you want to get in the pool? Go ahead, jump in!"

So much water, clean, warm water. She does want to swim in the pool, oh, she does — but not like this. "Thank you, not now."

***

She and the twins... with time they became as close to friends as it was earthly possible. There was that feeling, like being on the same page, more, standing side by side before a work of art only the three of them could fully admire and appreciate. They had get-togethers that they called "brainstorming in our think bowl." With time the twins became decent experts in her field but they still communicated in metaphors. When Leni discovered that her noncoding RNAs were secreted in *exosomes* — tiny bubbles that traveled the bloodstream from one bodily organ to another — from brain to gut, from liver to heart — Yric called them messages in a bottle, and sang her a few lines from an old song by the Police, in a startlingly true voice. It was one of those moments when it seemed to her that... And she pored over *A Message in a Bottle* lyrics like a teenage girl, as if some romantic clue, some hidden thrill was contained there... Ah, embarrassing to recall. Stuff it, lock it, throw away the key. Nothing ever happened. Years went by. She'd never ever even seen Yric by himself. Always with Paul. She wondered why the brothers would not marry — maybe it was the *twin thing*. The tie that feels freaky to people who do not know how it is to co-exist with an identical copy of yourself. Maybe she'd always feared to even *go there* because there was always this thought: what if she had to date them both?

***

Rollie again, between backstrokes: "So that was in 2014, the conference? And when did they build you an institute, again?"

"In 2017. It wasn't just for me."

"But you were the director."

"A deputy director. I was never good at anything other than doing bench science."

"I hate fake modesty!" This is bellowed amid puffs and spatters: Rollie is doing a butterfly stroke. "Sick of those clowns on the Senate committee —"

This she knows, at least, this she is used to. Rollie calls her up when he is feeling conflicted, and her part is to offer — therapy, of sorts. She'd like to think she is his tether. She only wishes her stomach would stop hurting.

"— If a fucking law is not applied at all or is practiced with a big fat *discretion* it should just — disappear from the books! Poof! Gone!"

Inwardly, Lenora laments how the words *Senate committee* don't even mean anymore what they used to mean in her day. She overhears more cussing from Rollie, words drowning in burbling and splashing water. "Rage! … Ugh, could just strangle the man! … Every one of these idiots has their heads screwed on straight — by me, no less — except this one fakey-modest bastard. Subpoena, my ass! You'll be wiping shit off the walls in an insane asylum with your subpoena! …"

Not an empty threat, she knows *that*. She has to interject, show that she cares. That's her job. She needs to start rerouting this conversation. But she fails to make herself do it, she wants to muffle her ears instead. All because of the news that Paul claims he's Yric.

***

Leni didn't want to leave academia. But being flush with Benes cash and having her lab stuffed with shiny new equipment when so many of her colleagues were holding theirs together with duct tape and chewing gum, hunting for hand-me-downs and scraping the bottoms of empty reagent vials — this was a recipe for strained relationships.

The Benes Institute for Epigenetic Inheritance grew and grew. The Benes Big Data farms plugged right into it, figuratively and literally. It was surreal, from Leni's vantage point. By then she achieved Step One of her program. She created a certain, artificial, of course, life experience for lab mice. Like this: give a mouse a series of zaps while vanilla scent was in the air. She demonstrated that this experience gave rise to a certain RNA in exosomes — a certain message-in-a-bottle, in Yric's words. This message meant something like (in Yric's words, again), *Life is bad if you smell vanilla! Full on stress!* and it was sent from the mouse's brain to all four corners of the mouse's body. Now it was time for Step Two.

June 14, 2019, a Giants' game. Leni threw a party for her whole lab, their plus-ones and kids. She booked a sky box. There was a smorgasbord of food, a rack of beers and wines, and a programmable ice cream machine that birthed multicolored baseball-shaped ice creams. The occasion? They had completed Step Two of the gold standard — intercepted a *Life-is-bad-if-vanilla* RNA message, pulled it from one bunch of mice, purified and amplified it, and sent it into the bloodstream of another bunch of mice. The study was blinded, controlled for every contingency. After several more months of work they un-blinded the sample IDs and there it was: the mice that had received the RNA had lost their peace with vanilla.

Midway through the game her phone rang. "Leni? We are next to you. Care to join us?"

She slipped out of her sky box and went into the one next to it, where she saw Yric and Paul, a bottle of champagne, a chocolate cake, three glasses, and a ginormous touchscreen/display, on which Paul drew a star-eyed smiley face, and then added beneath, *What's next?* After the toast, settled in a cushy arm-chair, Leni said, "Next? Find out if our LIBIV, *Life-is-bad-if-vanilla*, that is, RNA tweaks with the DNA of the sperm, of course. Maybe it just makes an epigenetic bookmark on the DNA. Maybe it actually induces a DNA double strand break. Maybe it even hitches a ride — in the sperm — to the next generation — " She floundered and stopped. Maybe it was because she'd used the word *sperm* and inadvertently looked at Yric. Maybe it was just the funny angle Yric sat at that made those insomniac's shadows under his eyes so deep and blue.

Paul said, "That's all well and good but we need to think bigger. You have one message — LIBIV. Shouldn't there be a whole library of them?"

She refocused. "Of course there is a whole library, but one is all I need for the gold standard of proof to work. One word that I can control and follow around experimentally. LIBIV is as good as any."

Outside, the whole stadium whooped as one at a homerun, and fireworks shot up, but the display in the twins' sky box stared steadily with nothing but the smiley face.

"Okay but yet bigger, Leni. Dream, please?"

"Well, in the grand scheme of things we could eventually explain evolution of instinctive behaviors. Explain how spiders weave their webs and honeybees build their hives, and human babies refuse to crawl on glass tables."

Yric meanwhile plopped giant pieces of chocolate cake on three plates and served them. He sat down and ate his portion, listening. Then he aimed a fork at Leni. "Bigger," he said, "as in informationally bigger. Leni, you have at your disposal the biggest data crunching outfit there ever was, courtesy of your humble servants the Benes Bros, and you still think like a liberal arts and crafts college professor, beg your pardon. Seriously. We have the capacity to collect, decode and assign function to hundreds of messages-in-a-bottle. To hundreds of LIBs, *Life-is-bads* and LIGs, *Life-is-goods*. We can catalog every significant experience, every life's lesson, every hard-knock and fist-pump, every *oops* and *woo-hoo* worth remembering and passing on. Why don't we do that in addition to whatever else you want to do? And in humans, Leni, not in rodents or honey bees, for that matter."

She considered it. It was so grand and sweeping, so beyond the nose-to-the-bench pace — yes, pedestrian pace — of her research. And it made Yric's eyes shine.

She got on board. That was the inception of the *LIGs and LIBs* project. A grand effort to identify exosomal RNAs that meant *Life is good* or *Life is bad* in human beings.

***

"Made you heady with all this success, no?" Suddenly, Rollie stands in front of her, dripping water. Hunched over her hurting stomach, Lenora sees his bare feet on the tile, the water that pools in grout lines. She just shakes her head.

"Is that when you began your human experimentation?" she hears Rollie asking.

Before she knows it she's jumped to her feet, she's shouting. "*Rolland, what the hell.*" She can't believe she's yelling into his amused face, "How can you say that, you know this is absolutely not true! You know perfectly well what I did, you have no right to distort it. I told you time and again: I collected serum samples from properly consented adults to isolate EXOSOMES!" She shakes her shrunken fist with every syllable as if it is a slogan, *EXO! SOMES!* How pathetic she must look right now, a little old lady spewing nerdy-talk — but she can't hold back this fury, this despair. "To isolate them and identify their RNA content! That's it! That's what LIGs and LIBs was, nothing more. None of that 'unprecedented human engineering' nonsense the media ran with. None! Just a small sample of blood, drawn for analysis. I did not tweak, inject, add, subtract, change, implant ANYTHING! I did NOT experiment on humans! … And for god's sake, cover yourself already, I don't have to stare at your junk!" She shoves a towel into Rollie's abdomen.

Rollie shakes his head as if there is water in his ear, steps back — but keeps, thank god, the towel at his navel, and is, thank heavens, wrapping it around his loins now. The next comes out with a touch of role-playing pretense. "Don't tell me what to do. *Mother.*"

"I am not —" she snaps, startling herself, and Rollie is already snickering — bigger and bigger laughs burst through his curving mouth one after another as if in ever more rapid bubbles. As if some kind of hysterical water is beginning to boil inside him.

"— telling you what to do," Lenora recovers. She laughs because she is distressed. Rollie looks at her almost admiringly. Suddenly he grabs her across the waist and hoists her body over his shoulder. He is a strong man and she is a diminishing senior, bones going feather-light. "Come on in, get wet, it'll do you good." She protests but it doesn't matter.

He giant-steps into the pool, releases her into the parting-then-colliding waves of his entry. The current drags Lenora to the other end of the pool. She anchors herself to the pool's edge with her forearm, wipes her face, sucks in air. Mutters shakily, "What the fuck." She does not fear for her life. She has her other, bigger, twin fears, as she calls them.

She fears transgression. By Rollie. And she fears that Rollie will pick up his father's vision and pervert it. These days, someone like him can find a way to force LIBs and LIGs on hundreds of unwilling, unsuspecting "volunteers."

She starts talking just so Rollie doesn't. He can dump her in his pool, fully clothed, but he can't make her stop talking. "I am doing fine, thank you. Same old, same old, our little trials and tribulations. Our school's corporate sponsor is mad at us because some kid put graffiti over their logos. Reads: *Ignore the humans around you.* Now the whole school has to suffer, the staff is fined, the lunch program is cut back. Kids have enough crap in their lives as it is. Cory's uncle got copper poisoning from the shitty knock-off jacks he uses for his brain plug-in. Beth's older sister signed up as a research subject for a cosmetics corporation, went in — and nobody's heard from her ever since. Ela-Jo's parents jumped on a new all-in-one utilities package, connectivity plus water. Didn't read the fine print. Guess what part of the package took priority, the internet of course. Now they are down to a half a gallon a day for all three of them, because the package is called *Communicating vessels*, the poor idiots did not know what it means, could've looked it up on their internet, but, hey — they don't use the internet to learn things. And as for myself? I can even water my petunias, thank you very much, Rolland… Rolland, I know you've done a lot for us, providing us with clean water and the rest, but… is it too much trouble to somehow communicate to our sponsor not to be so hard on the school?"

Rollie has switched off the current and is floating face-up at the other end, manifestly silent.

She continues, "I am sorry that I shouted at you, but you do know, Rolland, that I was, as they had put it, *gathering* joys too. Not just sorrows. The Gatherer of Sorrows was just a catchy name the media invented. If it bleeds it leads. The *Gatherer of Joys* does not have the same ring to it, does it. The *Gatherer of Sorrows* on the other hand is not only catchy but smacks of something subtly sociopathic. The truth, Rolland, and you know it, is that every person who donated blood for our project did so voluntarily, was fully informed, expected no personal gain or health benefit, and believed that in the long run this would move the science forward. We went to people who self-identified and were questionnaire-identified as living in a *Life is Good* state. To a newly-wed couple in their honeymoon. To a grandpa making friends with his first granddaughter. To a long-time cancer patient who has learned she is in full and lasting remission. We went to successful and accomplished and happy people. And — yes, we also went to those who thought they'd failed and made nothing of themselves. Who were in a *Life is Bad* state. To a fifty-five-year-old family man, fired from his job and now unemployable. To an artist who's never won any recognition for her work. To a man trapped in a woman's body. To a vet with PTSD… Do you know, Rolland, have I ever told you that I too donated a blood sample to my own project? As a *Life is Good* person?"

Silence is the answer, at first. Water reflections are playing on the ceiling, meditative. Then the still floating Rollie says, " 'Course you are my mother. Your egg, you the mom."

Did he hear a word of what she'd said just now? Rollie continues: "And yes you did experiment on humans — on my father and uncle. Don't deny it."

Another twinge of pain in her stomach. "It was their idea. I was against it but I could not stop them. They would have done it with me or without me."

"I don't really mind human experimentation," Rollie goes on, musingly. "Just not in the name of science. And not by the government."

This is the Rollie that scares the bejesus out of her. She feels tired and feeble-minded. She's fooling herself thinking she can — no, forget *fix*, just — slow him down, hold him back. Just look at him! And yet… he is the one who summons her here. He needs her. She begins to climb the stairs out of the pool, clenching the rails too hard. Water drains out of her clothes, making them cling, weighing her down. Rollie must be watching. She knows he won't come to help her up the stairs and hates to look like she needs help. When he gets like this she needs to look strong. She tries to at least sound strong. "Really, Roland. You ARE government."

"What? I am not." His voice is still coming from a pool's length away and sounds like he is yawning, or stretching. Yet there is also a cold, sharp edge to it when he adds, "A certain Robert LeFevre said, way back when, 'Government is a disease masquerading as its own cure.' I am the cure masquerading as government."

***

January 5, 2022. A lodge in Aspen, CO, fire crackling in a stone hearth. The twins had gone on snowboarding while Leni had called it quits, frostbitten and outmatched. She was sitting by the fire, wrapped in a plush blanket and sipping a steaming espresso with Bailey's when the twins stomped in — covered in snow, flushed, wind-blown. "We have an idea and a plan," Paul announced while Yric stepped out of his Gortex and toppled onto the couch so hard, his body bounced.

"Like in 2019 — remember? — you need a gold-standard proof. Right? A transfer of a message-in-a-bottle from human to human. Let's do it!"

By then the LIBs and LIGs project has yielded a handful of messages-in-a-bottle that had to mean *Life is Good,* and a few that had to belong to *Life is Bad* category. More so, Leni could now synthesize these messages in whatever quantities, and encapsulate them in exosomal bubbles just like the ones from which the original messages had come. She could feed these bubbles to human cultured cells and observe the effects. Yet Leni did not get it at first that Paul was talking about the LIBs and LIGs. She glanced at Yric, who had shored his eyes behind Smart-glasses. His lips moved as if he was reading to himself off the screen. "Yric?" she called. "Yes?" he said absent-mindedly.

"Does Paul really say you want human volunteers to be injected with LIB or LIG exosome preparations? Am I hearing it right?"

"Uh-huh."

She jumped to her feet, she paced. She shook her head, *no, no, no.* She insisted they were not yet in the right place to even consider a human trial of this sort.

They argued for a better part of the night. Paul appealed to her scientific curiosity and her high standards of proof. Paul said she was stonewalling. Then Yric said, how could she have the knowledge of something and refuse to apply it? Leni shouted that the very discussion was so wrong on so many levels; that she was fine with just knowing, not applying; that she was no goddamn Dr. Strangelove, she was a woman-scientist, and as a woman she'd never lose sight of a human aspect of any of it in pursuit of some brain-itching puerile cockamamie intellectual victory! Which words precipitated mutual accusations of sexism between her and Paul, then back-pedaling to square one, with a thickening air of sourness and irresolution. And then Yric said calmly, "If Paul and I volunteer as trial subjects, you cannot say no to us. Legally. Your work belongs to the Benes Institute."

This — and they had to have decided on it already — so utterly and completely knocked all polemic wind out of her, that a forlorn, all-encompassing "Why?" was the only word that escaped her throat.

"Intellectual curiosity," Yric said. She'd never forget his smile: half self-mockery, half a gentle reproach to her that there was something she'd never understand.

She should have resigned from the Benes Institute right then. But she felt a responsibility to stay and yes, an attachment to the work of her life. The design of the trial too, was Paul and Yric's. There was no placebo control, instead, one of the twins would receive a LIG into his bloodstream, the other — a LIB. They even devised protocols for their own regular physical and psychological evaluations. Leni's only input was that everyone involved had to be blinded as to which of the twins got what — that was to be revealed in five years after the start of the trial or if the trial was aborted early. The only way to avoid anticipation biases, she said, and the twins agreed to it.

A year into the trial, Leni heard that the twins were considering having kids. Making some kind of "arrangements" for that, rumors had it. She wasn't privy to the firsthand information now that they'd had their disagreement. For a while she thought 'the arrangements' meant nuptials, and it made her even more furious because she knew what the twins were after. The *gold-standard proof*. Transgenerational inheritance of the messages-in-a-bottle they were receiving in monthly injections. She sent them a 25-page manifesto on why this was risky and unethical but for all she knew, they went on with their plans.

And then Yric was found dead.

A suicide, but maybe a murder. There was an investigation, which ended with nothing. And then another kind of investigation, when the fact of the twins' experiment came into the light, dragging out on its tail the whole LIBs and LIGs project.

She grieved.

She resigned.

She testified in a hearing.

She did not say what Paul's lawyers were coaching her to say. So they cut her loose.

She never denied her responsibility. She felt guilty for letting the twins run away with her science.

She was nicknamed The Gatherer of Sorrows. The Benes Institute was vandalized. Her image was used to frighten little children with science and scientists and the research in human genetics in particular.

Paul Benes fought back. It was ironic, that his defense had become a foundation for the most restrictive policy ever toward studies in human biology. Paul's lawyers had managed to argue that all human DNA and all RNA — the metaphorical book and messages, law and its practice, all that *genome* and the many ways of its *expression* — was free speech, which the government had no right therefore to regulate, change, and improve directly or by sponsoring scientific research.

Only private citizens could do it — on their own accord and with themselves or with free and willing other private citizens as experimental subjects. Corporations were private citizens.

Ah, the changes this had precipitated. Not for the LIBs and LIGs, which had been scuttled, but elsewhere. Everywhere.

***

Lenora succeeds in climbing out of the pool, picks up some towels off the floor and swaddles herself. The damp towels hardly provide any warmth. Her hands, her chin are shaking. "I was against it," she repeats, staring at Rollie.

He stares back, idly treading water. The towel that has slipped off his middle is floating nearby, its corners spread out like for an embrace. "They played you like a fiddle," he says and blows bubbles. She shakes her head, "It's not like that."

He says, "So who got what? Who got a LIG and who — a LIB?"

The million-dollar *question,* asked again, asked at last. "Haven't I told you? The sample ID key was lost when the Institute was vandalized. So I do not *know* but my conclusion is Paul got a LIG, which means, you got a LIG. From Paul, your father. Assuming it ever worked, which is uncertain." She knows to look straight into Rollie's eyes when she says this. The angle must be funny though, because she suddenly sees traces of — she wants to say — Yric in him, even as she insists that there is absolutely no reason she should be thinking this way, because Yric and Paul had *absolutely* identical, clonal features. And then she finally *understands.* She understands Yric, and maybe Yric and Paul both. Maybe all they'd ever wanted was to become different from each other. To diverge, evolve. To become opposites rather than copies. To part like scissors' blades, to hate or love each other — just not to feel that innate, umbilical oneness. They had wanted it so badly that they were ready to absorb thousand-fold concentrated, weaponized sum-totals of other people's life's evolutions for that.

LIBs and LIGs.

If only she'd known. She could have told Yric they already *were* different, and she didn't need to be an epigeneticist to know that, that she'd always singled him out, that there was never a doubt in her mind from that very first meeting and the elevator ride, that Yric was Yric and Paul was Paul.

Had she said it. Had he listened. Had any of them known what the fallout of their personal quests would be.

She hears Rollie say something and it takes her awhile to tune back in because her mind has traveled so far, far away, but then comes the brutal reverse, the dreaded, "Explain to me then, *mother*, why do I feel like a monster?"

She frantically casts about for a reply. Then it comes, a straw. "Rolland, listen to me. Just hear me out, all right? Yes, I am your mother, but not because you come from my egg, or not so much because of it. Here is something you don't know yet. Back then, the Benes Institute had started a program for female employees: cryopreservation of eggs. As a deputy director, I had to lead the troops by example. So I did it. Now, when I resigned I was supposed to take my eggs with me. Unlike my science, they were my lawful property, but when I made the request, it appeared they had been lost or destroyed. I thought it happened during the break-ins. But then I started to suspect it was earlier than that. I managed to reach Paul and I confronted him. He confessed... To put it simply, he and Yric had stolen my eggs to... to make you. To tell you the truth... I was livid, I really was. But then Paul introduced me to you. You were four. Your surrogate mother who'd borne you, was not in the picture. And so we... made friends. And spent time together. We have our memories, you and I. Good memories. Remember our Sunday sandwich, the *sundwich*? And the bicycle fairy? *That* is why I am your mother."

Rollie silently heaves himself out of the water. He picks up, wraps around his waistline and tucks in a towel. He goes to the bar and pours himself a drink so haphazardly, half the bourbon tsunamies onto the counter.

"Can I have a dry towel, please?" Lenora says.

Rollie slides out a drawer, picks three up between his fingers and tosses them in Lenora's lap. She makes a cocoon around herself. Rollie drinks, silent, holding the glass near his mouth. "Stolen, huh." He studies her. "So what are you saying?"

Looking up at Rollie, Lenora says, "You said you are angry at someone in the Senate. That someone is making a stand against you. You are deciding on what to do about it... about that man, aren't you? Please don't do anything... irreversible. Anything that you will regret later. I am asking you as your mother. I was just as angry back then as you are now but I... did not act on my anger."

Rollie chuckles, shaking his head. "You old bird," he says almost softly.

Lenora takes it as a good sign.

She thinks — hopes — that this will be the end of it for today, and she can go home. But the next moment Rollie's stare stiffens, goes distant. He is thinking. Calculating. "*Stolen.* You could have sued. Could've pulled the rug straight from under daddy's feet. Right that very moment... Obviously you didn't... Hmm. He paid you off... He paid you off, didn't he?" He stares into Lenora's face trying to pry out an answer. "No. It was about custody. No? ...

Visitation rights. No? Okay. If you'd pressed charges, there would have been a maternity test... a test on me from which you would have learned that —

"— Christ, you could have run this test on me at any time! And maybe you did. You can run it now if I make you, and it will tell you which message-in-a-bottle had been there because you know where to look and what to expect —"

He is close. So close that her heart freezes like a little girl on the cold floor of her diaphragm as her stomach goes berserk below. She can barely take a breath, almost seeing how Rollie's mind spins faster and faster behind his eyes. "— unless... unless you already knew. You knew! There had been no break-ins at the Institute. You'd gone in after uncle's death and you unsealed the key to sample IDs... Didn't you? You are lying to me! My old man, daddy Paul, now says he is Yric. So tell me now, is he senile and deluded or senile and blabbering out the truth, at long last —"

So close.

She is so terrified. But she needs to look strong. Stronger. She can do it. She straightens her back, sits taller. She sees Rollie's mind, grinding yet faster, she realizes that it is darting up and down, right and left over that imaginary eight-way table of all possibilities that could have been, that still can be — he is discovering this very moment — the case. Rollie's face grows strangely pinched, he looks like a kid who is trying to give a right answer to his math teacher — and is failing. A LIG? Or a LIB for Paul. Paul? Or Yric as a survivor. A suicide? Or a fratricide. Whose son is he, and what does it mean? He is working, calculating all the scenarios and all their multiplying implications, trying to figure out his mother's *game*, her *angle, whatever she has on him,* because she *must* have something! But he will never — she can see it so clearly now — grasp the full truth. It will evade him because he's got no sensory organ for trust, or forgiveness, or penitence, no calculus for knowing everything there is to know and not taking advantage of it. Nor will he understand her science, he just doesn't have the education.

And this gives her hope. She breathes in and repeats her half-lie one more time, speaking slowly and distinctly, tip-toeing carefully between her two worst fears. Her twin fears. "Paul, your father, *must* have received a LIG, Rolland. It could not have happened any other way."

Rollie takes a swig of his bourbon. She thinks: he *can't* compel her to test him, because — well, because she is at that wonderful age when she'll keel over before she'd do anything she really does not want to do. She thinks: he is nowhere near restarting the LIGs and LIBs experimentation. Not yet. Still not. She relaxes a coil of her arms around her stomach. She repeats, "A LIG. Paul's life was good, Rolland. My life was good. What you feel, what burns inside you, the rage, the scorn — they are your responsibility, they are not forced upon you by your origin, by something that was done to you before you were even born. You should be a good man who does not hurt people."

Rollie grins. "And if not?"

"Then there is no transgenerational inheritance."

She frees herself from the towels and gets up. Gingerly, she approaches Rollie and wraps her arms around his waist, feeling how her son, her joy and sorrow, her *gold-standard proof*, tenses up, then relaxes. For a brief moment, she presses her cheek to his sternum, and his heart booms into her ear. Pulling away, she adds, "But you still have your parents. An old decrepit dad and an old bird mom; I know we are not much to speak of anymore but we are still around and we wish you well. And we are like that stretch of earth between you and the precipice where there is erosion. For as long as we hang on — you are away from the edge."

Rollie is gazing at the floor. His shoulders are slumped. She takes it as a good sign.

***

The pteroglider deposits Lenora in the yard of her school, as she asked. She's decided against being dropped off at her house — lest she'll find that vodka bottle and drink it up. The machine lifts and zooms away. She enters the school building, dark, and follows the nightlights to her classroom. She sits down at her desk, clasping her head between her palms. Something moves under the kids' desks.

"Miss Shelley?"

"Ela-Jo? What are you doing here this late?" Lenora flips the light switch. The girl is climbing out of a "fort" she'd made herself of desks and chairs. She snuffles. "I was just… waiting for you."

"Rough times at home, huh?"

Ela-Jo only jerks her shoulders. Her face squirms but she holds it together. Always a trooper.

Lenora goes to the sink and presses her thumb into a reader, then, when the little light turns green, she opens the faucet and fills up a glass. "Here."

"Where you been?" Ela-Jo says, gulping down the water.

"Ah. Doing my penance, young lady… Why don't you wash your hands and face, while you're at it. And your hair. There's enough water, I make sure of it." She searches in her pocket and pulls out a disintegrating damp paper towel with diffused purple streaks. "Oh my, I completely forgot. Our pH indicator. I guess we'll have to make a new one tomorrow, right?"

Ela-Jo curves her lips. "What for? We still can't fix the water."

Lenora sighs. "True. But if you know something… if you know it then at least some day… maybe… you can fix it." She wipes her eyes with the heel of her hand. "That's all."

Afterword

Let me just say it: every element of Dr. Mireles's science is real or rooted in reality, and refers to actual, developing research in epigenetics. As the name suggests, epigenetics is the study of inheritance by means above genetics, on top of genes and the stuff they are made of — the DNA molecule. But what does it mean, exactly?

That is a matter of incredible breakthroughs in molecular genetics of the 21st century. In the nineteen nineties the paradigm was largely like this: there is DNA. DNA is parceled into genes that code proteins. Genes are expressed, i.e. transcribed into RNA. RNAs are translated into proteins (unless they are of the two classes of RNA that service translation itself). Proteins make cells and service them. How is the work of this mechanism controlled and adjusted to various conditions? By proteins. Regulatory proteins sense changes in the environment, bind to DNA near or within the genes, and influence expression of these genes. To permanently change expression of a gene one therefore needs to mutate the DNA of the gene or of its regulatory protein.

This was a good start but it couldn't be all there was. First, in complex organisms like us (and other vertebrates) protein-coding genes occupy less than 2% of all genomic DNA our cells carry. Why so wasteful? Second, DNA in a cell is not naked but is elaborately wrapped in and around special proteins, seemingly making access to it by regulatory proteins more challenging than it needs to be. Why?

As it turns out, the circuitry DNA → RNA → protein → DNA is only one part of the system. A lot of DNA codes RNAs that never translate into proteins. But these so-called non-coding (nc) RNAs carry out regulatory functions on par with the proteins. In particular, they regulate gene expression. Moreover, this RNA-mediated regulation can be quite permanent, thus introducing to us one area epigenetics studies: changing gene expression not by mutating a gene but by making it a target of a long-term repression or activation by a ncRNA.

And what about those protein wrappers of DNA? Turns out, the most ubiquitous of them, called histones, offer their surfaces for putting semi-permanent chemical "tags" next to this or that gene in DNA, in order to either ease the access of regulatory proteins to the gene or prevent it altogether. Thus developed another brunch of epigenetics: a study of placement and erasure of these tags and their role in long-term, stable changes in gene expression that accumulate over the life time of an organism. In other words, both ncRNAs and histone tags are examples of adaptive changes that even two genetically identical organisms like Yric and Paul Benes can accumulate as they get older.

It only gets more interesting from here. Persistent presence of a repressive histone tag invites chemical modification of DNA itself, which further discourages regulatory proteins from binding it. This modification can degrade, causing a heritable mutation in DNA. This may be one way by which an epigenetic, adaptive change eventually forces genetic change. But that's not all. First, not only physical conditions such as starvation, but also mental conditions such as PTSD affect gene expression by epigenetic means. Second, ncRNAs and histone tags can cooperate in repressing genes, with ncRNA attracting repressive histone tags. Third, in addition to controlling gene expression, ncRNAs and histone tags also control proteins that repair broken or damaged DNA. And where there is repair — there is making of mistakes, i.e. making mutations. And lastly: yes, some ncRNAs are secreted from cells in exosomes and thus can in theory exert their effects not only in a cell of their origin but in some distant cell — even in a recipient organism. ncRNA content of exosomes is known to vary under different life conditions of an organism, and in health versus disease. With all this insight into the ways cells can accumulate and retain adaptive changes over an organism's life, researchers are setting their sights on the next question: can some of these changes be transferred to an organism's progeny? Experiments begin to say: yes.

Now we are at the edge of the known and are venturing into the hypothetical but not unlikely. What Lenora proposes is that a certain class of ncRNAs is can be transferred from cell to cell, organ to organ in exosomes. This includes going to the germline — sperm and egg cells. Inside a cell, these RNAs bind to the DNA of genes and alter their expression. If these ncRNAs are abundant or persistent, they also can: either stimulate a repressive histone tag and then a mutation at the site of their binding; or alter DNA repair so that a mutation is more likely to be introduced at the site. Lenora envisions some combination of both, in fact she thinks her ncRNA actually attracts repair proteins to the site of its binding, and induce these proteins to repair what is not broken — and introduce a mutation. Either way, if the DNA is broken, this increases the chance of mutation. Lenora knows the identity of her ncRNAs, knows which gene each binds, knows what histone tag or mutation footprint this can leave on or in DNA. If she also knows which ncRNA each twin took, she can not only distinguish Yric from Paul, but also figure out which one is Rolland's father — assuming transgenerational transfer did happen. What neither she nor the twins know at the start of the trial, is what systemic physiological and behavioral effects massive bombardment by LIG and LIB ncRNA will produce in the recipients and their progeny.

The rest, as they say, is fiction. The experiment is ongoing.

For a good recent overview of transgenerational inheritance studies see for example *Epigenetics: The sins of the father, by Virginia Hughes. Nature News 2014, v.507, pp.22–24*